Thermal Physics Tutorial with Python Simulations

This book provides an accessible introduction to thermal physics with computational approaches that complement the traditional mathematical treatments of classical thermodynamics and statistical mechanics. It guides readers through visualizations and simulations in the Python programming language, helping them to develop their own technical computing skills (including numerical and symbolic calculations, optimizations, recursive operations, and visualizations). Python is a highly readable and practical programming language, making this book appropriate for students without extensive programming experience.

This book may serve as a thermal physics textbook for a semester-long undergraduate thermal physics course or may be used as a tutorial on scientific computing with focused examples from thermal physics. This book will also appeal to engineering students studying intermediate-level thermodynamics as well as computer science students looking to understand how to apply their computer programming skills to science.

Series in Computational Physics

Series Editors: Steven A. Gottlieb and Rubin H. Landau

Parallel Science and Engineering Applications: The Charm++ Approach
Laxmikant V. Kale, Abhinav Bhatele

Introduction to Numerical Programming:A Practical Guide for Scientists and Engineers Using Python and C/C++
Titus A. Beu

Computational Problems for Physics: With Guided Solutions Using Python
Rubin H. Landau, Manual José Páez

Introduction to Python for Science and Engineering
David J. Pine

Thermal Physics Tutorial with Python Solutions
Minjoon Kouh and Taejoon Kouh

For more information about this series, please visit: https://www.crcpress.com/ Series-in-Computational-Physics/book-series/CRCSERCOMPHY

Thermal Physics Tutorial with Python Simulations

Minjoon Kouh and Taejoon Kouh

CRC Press
Taylor & Francis Group
Boca Raton London New York

CRC Press is an imprint of the
Taylor & Francis Group, an **informa** business

First edition published 2023
by CRC Press
6000 Broken Sound Parkway NW, Suite 300, Boca Raton, FL 33487-2742

and by CRC Press
4 Park Square, Milton Park, Abingdon, Oxon, OX14 4RN

CRC Press is an imprint of Taylor & Francis Group, LLC

ISBN: 978-1-032-25756-3 (hbk)
ISBN: 978-1-032-26343-4 (pbk)
ISBN: 978-1-003-28784-1 (ebk)

DOI: 10.1201/9781003287841

Typeset in SFRM font
by KnowledgeWorks Global Ltd.

Publisher's note: This book has been prepared from camera-ready copy provided by the authors.

To our parents, Yong Woo and Byung Ok Kouh

Contents

Preface

"Doing" physics is an integrative and multi-modal activity where one thinks about natural phenomena with many different intellectual tools and approaches. Mathematics is one of the most powerful and essential tools of a physicist, or may even be considered as the language of physics. However, in recent years, computational methods have risen to complement and supplement the traditional, mathematical approaches to physics. As the new generation of physicists is expected to be well versed in modern computational tools, this tutorial was written with the goal of introducing a few elementary skills in data visualization, modeling, and simulation with a popular (as of the 2020s) programming language, Python, within the context of classical thermodynamics and statistical physics.

This book provides step-by-step instructions for each of the programming examples, and prior experience with Python is not necessary. If you are just venturing into the world of Python, the official homepage of the Python language (`www.python.org`) is a great place to visit. There are other resources on Python, many of which are free and easily accessible online. There are different ways to set up your own computing environment, so that you can follow the codes in this book. For example, you may download and install the Anaconda distribution, which contains an interactive Jupyter Notebook environment as well as key Python modules. You may also use a cloud-based Python environment like Google Colab. See Appendix for more information.

Several popular topics from classical thermodynamics are covered in Chapters 2, 3, and 4. The premise of modern statistical mechanics is introduced in Chapters 5 and 6. Chapters 7 and 8 discuss the connection between classical thermodynamics and statistical mechanics in the context of an ideal gas. The next chapters introduce other examples of a thermal system (two-state system, simple harmonic oscillator, and

Einstein solid). The final chapter is about random and guided walks, a topic that is independent of earlier chapters and provides a glimpse of other areas of thermal physics.

ABOUT THE AUTHORS

T. Kouh earned his B.A. in physics from Boston University and Sc.M. and Ph.D. degrees in physics from Brown University. After his study in Providence, RI, he returned to Boston, MA, and worked as a postdoctoral research associate in the Department of Aerospace and Mechanical Engineering at Boston University. He is a full faculty member in the Department of Nano and Electronic Physics at Kookmin University in Seoul, Korea, teaching and supervising undergraduate and graduate students. His current research involves the dynamics of nanoelectromechanical systems and the development of fast and reliable transduction methods and innovative applications based on tiny motion.

M. Kouh holds Ph.D. and B.S. degrees in physics from MIT and M.A. from UC Berkeley. He completed a postdoctoral research fellowship at the Salk Institute for Biological Studies in La Jolla, CA. His research includes computational modeling of the primate visual cortex, information-theoretic analysis of neural responses, machine learning, and pedagogical innovations in undergraduate science education. He taught more than 30 distinct types of courses at Drew University (Madison, NJ), including two study-abroad programs. His professional experiences include a role as a program scientist for a philanthropic initiative, a data scientist at a healthcare AI startup, and an IT consultant at a software company.

ACKNOWLEDGEMENT

T. Kouh would like to thank the faculty members in the Department of Nano and Electronic Physics at Kookmin University for their support, along with his current and former students from the lab. Stimulating and engaging discussions with his students on various topics in physics have started him to mull over intriguing and entertaining ways of answering the questions, which has been valuable to completing this book. He is also grateful to his mentors, Prof. J. Valles from Brown University and Prof. K. Ekinci from Boston University, for guiding him through his academic career and showing him the fun of doing physics. Last but not least, his biggest and deepest thanks go to his dearest Sarah.

M. Kouh would like to thank his colleagues from the Drew University Physics Department, R. Fenstermacher, J. Supplee, D. McGee, R. Murawski, and B. Larson, as well as his students. They helped him to think deeply about physics from the first principles and from different perspectives. He is indebted to his academic mentors, T. Poggio and T. Sharpee, who have shown the power and broad applicability of computational approaches, especially when they are thoughtfully combined with mathematical and experimental approaches. His family is a constant source of his inspiration and energy. Thank you, Yumi, Chris, and Cailyn!

Calculating π

This chapter introduces a few key programming terms and concepts, while we do a warm-up exercise of numerically estimating π. This important irrational number, which begins with $3.141592\cdots$, appears in many mathematical contexts and can be estimated in different ways. For example, with clever geometry, one could approximate a circle as a fine polygon and calculate the ratio of the circumference and the diameter of a circle, since circumference over the diameter of a circle is $\frac{2\pi r}{2r} = \pi$. Alternatively, one could use an infinite series that converges to π.[†]

Perhaps one of the easiest ways to obtain the value of π is to use the fact that $\cos\pi = -1$, which means $\cos^{-1}(-1) = \arccos(-1) = \pi$. The following lines of code print out the value of the inverse cosine function, which is equal to π. The **import** command in Python expands the functionality of the calling script by "importing" or giving access to other modules or libraries. Here, the **import math** command imports the mathematical library, **math**, which contains many predefined mathematical functions and formulas, such as arccosine **acos()**.

```
# Code Block 1.1

import math
print(math.acos(-1))
```

3.141592653589793

[†]See "The Discovery That Transformed Pi," by Veritasium, at www.youtube.com/watch?v=gMlf1ELvRzc about these approaches.

Let us introduce other programming syntax and useful numerical tools. The next block of code imports two other modules we will rely on throughout this book. The first line (`import numpy`) allows us to efficiently work with vectors, matrices, or multidimensional arrays. `numpy`, or NumPy, stands for Numerical Python. The phrase `as np` allows us to use the imported `numpy` module with a short-to-type and easy-to-remember name, `np`. Similarly, the second line (`import matplotlib.pyplot as plt`) allows us to easily make plots and data visualizations in our code. Most of the code examples in this book will almost always start with these two `import` statements.

```
# Code Block 1.2

import numpy as np
import matplotlib.pyplot as plt
```

Let's start with a few straightforward plots. Try to decipher what each line of code is doing. You can run the code without or with a particular line by adding or removing `#` at the beginning of the line.

The command `plt.plot((-1,1),(1,1),color='black', linewidth=5)` draws a black horizontal line that connects two points at (-1,1) and (1,1) – top side of the square – and the rest of the sides are similarly added in the subsequent commands with `plt.plot()`. `plt.xlim()` and `plt.ylim()` which help to set the limits of x- and y-axes. `plt.axis()` controls the axis properties of a plot based on the argument within the command. For example, by using `equal` with `plt.axis()`, we can generate a plot with an equal scale along x- and y- axes, and the argument `off` will hide the axes. Finally, `plt.savefig()` and `plt.show()` allow us to save and display the resulting plot.

```
# Code Block 1.3
# Make a square with a side of 2.
plt.plot((-1,1),(1,1),color='black',linewidth=5)
plt.plot((1,1),(1,-1),color='gray',linewidth=10)
plt.plot((1,-1),(-1,-1),color='black',linewidth=5,linestyle='dashed')
plt.plot((-1,-1),(-1,1),color='gray',linewidth=5,linestyle='dotted')
plt.xlim((-2,2))
plt.ylim((-2,2))
plt.axis('equal')
plt.axis('off')
plt.savefig('fig_ch1_box.eps')
plt.show()
```

Figure 1.1

1.1 ESTIMATING π WITH A POLYGON

To make a plot of an approximate circle, let's create an array of angle θ values between 0 and 2π radians in small steps of $\Delta\theta$. Then, the coordinates of a unit circle (with radius $= 1$) are given by: $(x, y) = (\cos\theta, \sin\theta)$. We can connect these successive points with short lines that collectively approximate a circle.

```
# Code Block 1.4

pi = 3.141592
delta_theta = 0.4
theta = np.arange(0,2*pi+delta_theta,delta_theta)
x = np.cos(theta)
y = np.sin(theta)

N = len(theta)
print("Number of data points = %d"%N)
for i in range(N-1):
    plt.plot((x[i],x[i+1]),(y[i],y[i+1]),color='black')

# Connect the last point to the first point.
plt.plot((x[-1],x[0]),(y[-1],y[0]),color='black')
```

```
# Put a box around the polygon.
plt.plot((-1,1),(1,1),color='gray')
plt.plot((1,1),(1,-1),color='gray')
plt.plot((1,-1),(-1,-1),color='gray')
plt.plot((-1,-1),(-1,1),color='gray')
plt.xlim((-2,2))
plt.ylim((-2,2))
plt.axis('equal')
plt.axis('off')
plt.savefig('fig_ch1_draw_circle.eps')
plt.show()
```

```
Number of data points = 17
```

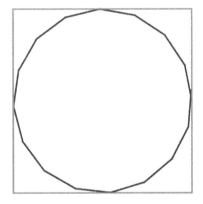

Figure 1.2

Let's go over a few key lines in the above code. The equal sign = assigns a value on the right to a variable on the left. Hence, **pi = 3.141592** means that the variable named **pi** is assigned to a value of 3.141592. Similarly, the variable **delta_theta** is assigned to a value of 0.4 (radian), which can be made smaller to create a even finer polygon.

There is a lot going on with **theta = np.arange(0, 2*pi+delta_theta,delta_theta)**. Here, we are creating an array or a vector of numbers and assigning it to a variable named **theta**. This array is created with the **numpy** module imported earlier. Since we have imported **numpy** with a nickname **np**, we can access a very useful function `arange()` within the `numpy` module as `np.arange()`. `arange()` creates an array of numbers specified by three numbers within the parentheses. The first number gives the starting point (inclusive) of the range, the second number gives the ending point (exclusive), and the third number gives the spacing between successive numbers. For example, `np.arange(0,1,0.2)` produces an array with numbers 0, 0.2, 0.4, 0.6, and 0.8. A similar function `range()` is also used quite often, and it does not require the `numpy` module, so `range()` can be called without adding `np.` in front of it. It returns a sequence of numbers starting from zero with an increment of 1 by default. `range(5)` returns 0, 1, 2, 3, and 4.

The next two lines `x = np.cos(theta)` and `y = np.sin(theta)` perform the mathematical operation of calculating $\cos\theta$ and $\sin\theta$ on an array of angle values, `theta`, using the `cos()` and `sin()` functions within the `numpy` module. Hence, the resulting variables, `x` and `y`, are also an array of numbers.

`len()` returns the length of an array, so `N = len(theta)` assigns the number of elements in the `theta` array to a variable `N`. This is useful, because in the next few lines, we will be plotting a line between two successive points in the polygon. `for i in range(N-1):` is called a forloop, which iterates or loops over the same operation within its indented block while the variable `i` goes from 0 (inclusive) to `N-1` (exclusive). We access each element of the array `x` and `y`, using `i` as an index. For example, `x[i]` refers to the `i`-th element of array `x`. `x[i+1]` refers to the next element. Therefore, `plt.plot((x[i],x[i+1]),(y[i],y[i+1]))` draws a line between the `i`-th and `i+1`-th elements.

The first element of an array is indexed with zero, not one. The second element is indexed with one, and so forth. The last element of array `x` is then accessed as `x[N-1]` (not `x[N]`), and it can also be referred to as `x[-1]`, which is useful if the length of the array is not known. The second to the last element is `x[-2]`. `plt.plot((x[-1],x[0]),(y[-1],y[0]))` puts a line between the last and the first coordinates.

Now, we have a polygon that approximates a circle. We can estimate the value of π by calculating the total distance around the polygon, which should approach the circumference of a circle, or $2\pi r$.

```
# Code Block 1.5

# Calculate the perimeter of a polygon,
# by adding up the distance between two successive points.
d = 0 # Start with zero.
for i in range(N-1):
    # Keep adding the distance between the next two points.
    d = d + np.sqrt((x[i]-x[i+1])**2 + (y[i]-y[i+1])**2)
# Finish up by adding the distance between the first and last point.
d = d + np.sqrt((x[-1]-x[0])**2 + (y[-1]-y[0])**2)

pi_estimate = d/2 # because d = 2*pi*r with r=1 for a unit circle
print('Our estimate of pi (with a polygon of %d sides)'%N)
print(pi_estimate)
```

```
Our estimate of pi (with a polygon of 17 sides)
3.237083436148559
```

In the above block of code, we calculated the Euclidean distance between two successive points, $\sqrt{(x_i - x_{i+1})^2 + (y_i - y_{i+1})^2}$, and then kept track of the total using the variable d inside a **for**-loop. In order to obtain an even better estimate, try decreasing the value of **delta_theta** and re-running the code blocks from the top.

1.2 ESTIMATING π WITH RANDOM DOTS

Here is another approach for estimating π, using the fact that the area of a circle is πr^2. Let's consider a square with a side of 2 and an inscribed circle with a radius r of 1. The square would have an area of 4, and the area of the circle would be π. The ratio of the areas of the circle and the square would be $\pi/4$. To estimate the ratio of these areas, we will generate a large number of random dots within the square, as if we are sprinkling salt or pepper uniformly over a plate. The position of these dots will be generated from a uniform distribution. Then, we could compare the number of dots inside the inscribed circle and the square.*

---

*We might imagine doing this simulation in real life by throwing many darts and comparing the number of darts that landed inside of a circle. The precision of estimation will increase with the number of darts, but who would do something like that? See "Calculating Pi with Darts" on the Physics Girl YouTube Channel (www.youtube.com/watch?v=M34T071SKGk).

```
# Code Block 1.6

# Generate random numbers between 0 and 1 with np.random.rand().
# Stretch their range by multplying by 2 and subtracting by 1,
# so that all dots fit within the square with a side of 2.

N = 5000 # number of points.
x = np.random.rand(N)*2-1
y = np.random.rand(N)*2-1

plt.scatter(x,y,s=5,color='black')
plt.xlim((-2,2))
plt.ylim((-2,2))
plt.axis('equal')
plt.axis('off')
plt.savefig('fig_ch1_random_num.eps')
plt.show()
```

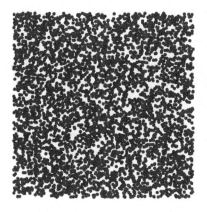

Figure 1.3

In the above code block, a large number (N = 5000) of random values
between 0 and 1 are generated by np.random.rand() function. The
range of these values is easily manipulated by multiplying them by two
and then subtracting them by 1, so we can create an array of values
between −1 and 1.

In the following code block, the distance of each point from the origin is calculated by `dist = np.sqrt(x**2+y**2)`, so that `dist` is also an array. The next line in the code, `is_inside = dist < 1`, compares the distance with a constant 1. If distance is indeed less than 1, the comparison `< 1` is true (or a boolean value of 1), and the point lies within a circle. If distance is not less than 1, the comparison `< 1` is false (or a boolean value of 0). Therefore, the variable `is_inside` is an array of boolean values (true/false or 1/0) that indicate whether each point lies inside a circle or not. Finally, by summing the values of this array with `np.sum(is_inside)`, we can calculate the number of points inside the unit circle.

We also note a clever way of selectively indexing an array. `x[dist<1]` returns a subset of array `x` which meets the condition where `dist<1`. In other words, it returns the `x` coordinates of the points that lie within the circle. Hence, `plt.scatter(x[dist<1],y[dist<1],color='black')` makes a scatter plot of the points in the circle with a black marker.

Sometimes it is convenient to package several lines of code into a function that accepts input arguments and returns an output. In the following code block, we created a custom function `estimate_pi()`, which takes an optional input argument `N` with a default value of 500. This function calculates an estimate of π using `N` random points.

```python
# Code Block 1.7

# Calculate the distance of each point from the origin.
# If the distance is less than 1, the point is inside of the circle.
dist = np.sqrt(x**2+y**2)
is_inside = dist < 1
N_inside = np.sum(is_inside)

# Estimate pi based on the ratio of number of dots.
pi_estimate = 4*N_inside/N
print('Our estimate of pi (with %d random dots)'%N)
print(pi_estimate)

plt.scatter(x[dist<1],y[dist<1],s=5,c='black')
plt.scatter(x[dist>1],y[dist>1],s=5,c='#CCCCCC')
plt.xlim((-2,2))
plt.ylim((-2,2))
plt.axis('equal')
plt.axis('off')
plt.savefig('fig_ch1_pi_circle_square.eps')
plt.show()
```

```
Our estimate of pi (with 5000 random dots)
3.1256
```

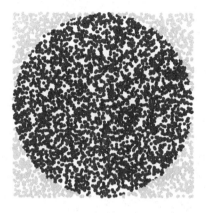

Figure 1.4

```
# Code Block 1.8

# Let's define a function that estimates pi for different N.
def estimate_pi (N=500):
    xy = np.random.rand(N,2)*2-1
    dist = np.sqrt(np.sum(xy**2,axis=1))
    pi_estimate = 4*np.sum(dist<1)/N
    return pi_estimate
```

We will use our custom function, **estimate_pi()** to investigate how good our estiamte of π is for a different number of points **N**. We will try **N** of 100, 500, 1000, 5000, and 10000. Because we are choosing **N** random values at each run, we will make an estimate of π 30 times (**N_trial**) for each **N**.

The last code block illustrates how to work with a two-dimensional array. The array **result** stores the result of each simulation. It is initialized as an array of zeros with 30 (**N_trial**) rows and 5 (**len(N_range)**) columns. Its content is updated with the command **result[trial,i] = pi_estimate** inside of the nested **for**-loops which independently increment the indices, **trial** and **i**. This code also uses **enumerate()**, a built-in Python function, which loops through the elements of an array with a corresponding index value. The final plot shows, as expected, that the estimates are more consistent, or the spread of the estimates is smaller, with higher N.

```
# Code Block 1.9

N_range = [100,500,1000,5000,10000]
N_trial = 30
result = np.zeros((N_trial,len(N_range)))
for i, N in enumerate(N_range):
    for trial in range(N_trial):
        pi_estimate = estimate_pi(N)
        result[trial,i] = pi_estimate
    plt.scatter(i+np.zeros(N_trial)+1,result[:,i],color='gray')
# Overlay a box plot (also known as a whisker plot).
plt.boxplot(result)
plt.xticks(ticks=np.arange(len(N_range))+1,labels=N_range)
plt.xlabel('N')
plt.ylabel('Estimate of $\pi$')
plt.savefig('fig_ch1_boxplot.eps')
plt.show()
```

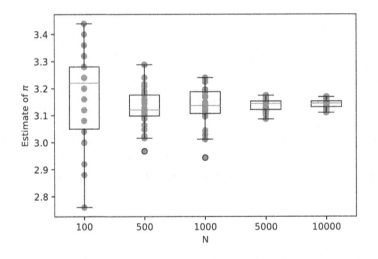

Figure 1.5

I

Classical Thermodynamics

## CHAPTER 2

Kinetic Theory of Gas

The idea that gas is composed of many tiny moving particles is obvious to us now, but it took many centuries of scientific inquiries for this idea to be accepted as a verified theory. We call this idea the "kinetic theory of gas." It was a significant breakthrough in physics, bridging two different domains of knowledge: classical mechanics, which usually deals with the force, momentum, and kinetic energy of an individual particle, and thermodynamics, which deals with the pressure, volume, and temperature of a gas.

Critical insights from the kinetic theory of gas are the mechanical interpretation of temperature and the derivation of the ideal gas law, $PV = nRT$, where P is pressure, V is volume, T is temperature, and nR is related to the quantity of the gas. More specifically, n is the number of moles of gas, and 1 mole is equivalent to 6.02×10^{23} particles (this quantity is known as Avogadro's number N_A). The proportionality constant R is known as a universal gas constant. Sometimes, the ideal gas law is also written as $PV = NkT$, where N is the number of particles and k is known as the Boltzmann constant. Therefore, $n = N/N_A$ and $N_A = R/k$.

As we will show in this chapter, the pressure of an ideal gas is a macroscopic manifestation of numerous microscopic collisions of gas particles with the containing walls. The gas temperature is directly related to the average kinetic energy of the constituent particles. As an analogy, consider the economy of a market, which consists of many individual transactions of people buying and selling products. These financial transactions collectively determine the macroscopic condition of a market, which an economist might even describe as being "hot" or "cold."

2.1 GETTING STARTED

Let's consider a simple one-dimensional, up-or-down motion, where a particle starts at the initial position of $y_0 = 0.5$ with a constant velocity of $v = -0.1$. The negative velocity indicates that the particle is moving downward. The position of the particle with a constant velocity at different times can be described as $y(t) = v \cdot t + y_0$, assuming no external force. In the following code, this expression is coded as `y = v*t + y0`, and it is the most important line.

When the particle hits a rigid wall at the position of $y = 0$ at the time of $t = 5$, it bounces off without any loss in energy. This process can be expressed mathematically by taking an absolute value of the particle's position, $|y(t)|$, since the position of this particle confined by the wall at $y = 0$ cannot be negative. The particle would now have a positive velocity of 0.1. As a result, the plot of y versus t has a V-shape.

```
# Code Block 2.1

import numpy as np
import matplotlib.pyplot as plt

# Calculate position at different times.
t = np.arange(0,10,0.1)
v = -0.1 # constant velocity
y0 = 0.5 # initial position
y = v*t + y0
y = np.abs(y) # Take an absolute value.

plt.plot(t,y,color='black')
plt.xlabel('Time')
plt.ylabel('Position')
plt.savefig('fig_ch2_y_vs_t.eps')
plt.show()
```

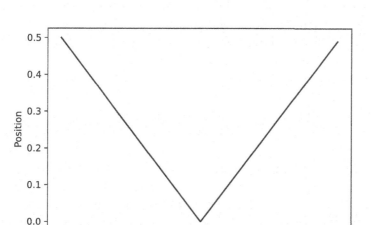

Figure 2.1

Now let's put this code into a more versatile and reusable format, a function that can be called with different parameters. Such a function is implemented below. We will update the position of a particle with a small increment dt, so that $y(t + dt) = y(t) + v \cdot dt$. In other words, after short time later, the particle has moved from the old position $y(t)$ to a new position $y(t + dt)$ by a displacement of $v \cdot dt$.

In the following code, we use a **for**-loop structure to update the position of a particle over time from **tmin** to **tmax** in a small increment of **dt**. The update starts from the initial position **y0**. The incremental update is accomplished by **current_y = y[i] + current_v*dt**. Note that, by using **time_range[1:]** in this **for**-loop calculation, we are just considering the elements in the **time_range** array excluding the first element (**tmin**), since the update is only needed for the subsequent times.

When a particle runs into a wall, the particle bounces off without a change in its speed or without loss of its kinetic energy. Only the sign of its velocity flips. In the code, this process is implemented with a statement, **current_v = -current_v**. The first **if**-statement handles the collision event with the bottom wall located at **ymin**, and the second **if**-statement is for the collision with the top wall at **ymax**.

When the particle bounces off of a wall at the origin ($y = 0$), simply taking the absolute value of the position prohibits negative position values and disallows the particle from punching through the wall. However, if we were to simulate a collision with a wall placed somewhere other than the origin, the particle's position would need to be updated more carefully. When the particle has just hit the bottom wall with a negative velocity (`current_v` < 0), the calculation of `current_y` = `y[i]` + `current_v*dt` would yield a value that is less than `ymin`. Therefore, the current position of the particle needs to be corrected by `current_y` = `ymin` + `(ymin-current_y)`. This command correctly calculates the position of the bounced particle to be above the bottom wall by the distance of `(ymin-current_y)`. When the particle hits the top wall at `ymax`, a similar calculation of `current_y` = `ymax` - `(current_y-ymax)` correctly puts the bounced particle below the top wall. We also keep track of the number of bounces by incrementing `Nbounce` by one within each `if`-statement.

```python
# Code Block 2.2

def calculate_position (y0,v,ymin=0,ymax=1,
                        dt=0.01,tmin=0,tmax=10,plot=False):
    # ymin and ymax are the boundaries of motion (walls).
    current_v = v
    time_range = np.arange(tmin,tmax,dt)
    y = np.zeros(len(time_range))
    y[0] = y0

    Nbounce = 0
    for i, t in enumerate(time_range[1:]):
        current_y = y[i] + current_v*dt # Update position.
        if current_y <= ymin:
            # if the particle hits the bottom wall.
            current_v = -current_v # velocity changes the sign.
            current_y = ymin + (ymin - current_y)
            Nbounce = Nbounce+1
        if current_y >= ymax:
            # if the particle hits the top wall.
            current_v = -current_v # velocity changes the sign.
            current_y = ymax - (current_y - ymax)
            Nbounce = Nbounce+1
        y[i+1] = current_y
    if (plot):
        plt.plot(time_range,y,color='black')
        plt.xlabel('Time')
        plt.ylabel('Position')
```

```
        plt.savefig('fig_ch2_bounce.eps')
        plt.show()
    return y, time_range, Nbounce

# Retrieve the returning values of y and Nbounce.
# The underscore symbol, _, ignores the returned value.
y, _, Nbounce = calculate_position(0.5,0.2,dt=0.1,tmax=30,
                                    plot=True)
print("Number of collisions with the wall: ", Nbounce)
```

```
Number of collisions with the wall:  6
```

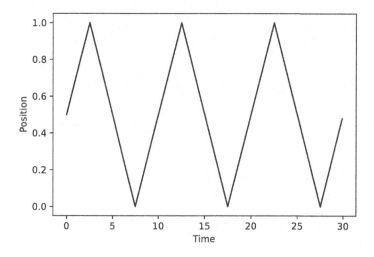

Figure 2.2

We can build on the above set of codes to simulate the motion of multiple particles. In the following code block, there is a **for**-loop that accounts for N particles with random velocities and initial positions. **np.random.randn(N)** generates N random numbers taken from a normal distribution with a mean of 0 and standard deviation of 1. Hence, multiplying it by 0.5 creates random numbers with a smaller variation. (There is a caveat. The velocities of gas particles are not normally distributed but follow the Maxwell-Boltzmann distribution. However, the main ideas discussed in this chapter hold up regardless of the distribution type. We will revisit this topic in a later chapter.)

A series of timestamps between the minimum and maximum time in steps of **dt** can be created with **t = np.arange(tmin,tmax,dt)**. The positions of N particles across this range of time are stored in a matrix named **pos**, which has the dimension of $N \times T$, where T is the number of timestamps, **len(t)**.

```
# Code Block 2.3

# Multiple particles.
N = 30
tmin = 0
tmax = 10
dt = 0.1
t = np.arange(tmin,tmax,dt)
pos = np.zeros((N,len(t))) # initialize the matrix.
Nbounce = np.zeros(N)

v = np.random.randn(N)*0.5
y0 = np.random.rand(N)
for i in range(N):
    # pos[i,:] references the i-th row of the array, pos.
    # That is the position of i-th particle at all timestamps.
    pos[i,:], _, Nbounce[i] = calculate_position(y0[i],v[i],dt=dt,
                                          tmin=tmin,tmax=tmax)

plt.hist(v,color='black')
plt.xlabel('Velocity (m/sec)')
plt.ylabel('Number of Particles')
plt.title("Initial velocities (randomly chosen).")
plt.savefig('fig_ch2_v0_distrib.eps')
plt.show()

for i in range(N):
    plt.plot(t,pos[i,:],color='gray')
plt.xlabel('Time (sec)')
plt.ylabel('Position (m)')
plt.ylim((-0.1,1.1))
plt.title('Position of $N$ particles versus time')
plt.savefig('fig_ch2_Nbounces.eps')
plt.show()

print("Do faster particles hit the walls more often?  Yes.")
plt.scatter(np.abs(v),Nbounce,color='gray')
plt.xlabel('Speed (m/sec) = |v|')
plt.ylabel('Number of Bounces')
plt.savefig('fig_ch2_bounce_vs_speed.eps')
plt.show()
```

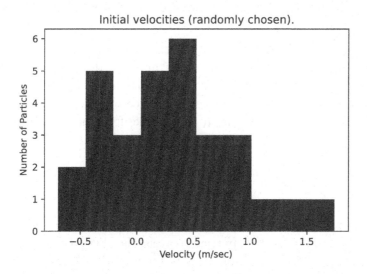

Figure 2.3

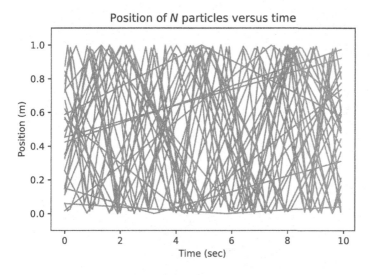

Figure 2.4

Although the numbers we are using for the above simulations are not particularly tied to any units, we will assume that they have the standard units of meters for position, seconds for time, and m/sec for velocity. We will label each axis with appropriate units as a way of illustrating good graphing practice.

The last graph from the above N-particle simulation shows a linear trend between the number of bounces and speed. This relationship comes from the simple fact that a faster particle would move between the two walls more rapidly and hence bounce off these walls more often. We can calculate how long it would take for a particle to make a roundtrip:

$$\Delta t = \frac{2L}{|v|},$$

where L is the distance between the walls (or **ymax-ymin**) and v is the velocity of a particle. The particle travels the distance of $2L$ before hitting the same wall. The frequency of bounce is $1/\Delta t$, and hence, the number of bounces is linearly proportional to $|v|$, as Figure 2.5 shows.

```
Do faster particles hit the walls more often?  Yes.
```

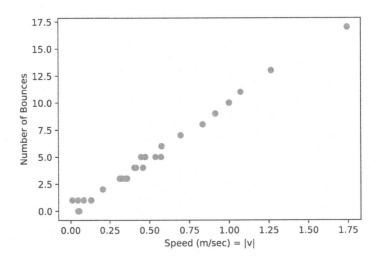

Figure 2.5

2.2 DERIVATION OF THE IDEAL GAS LAW

Now let's visualize a superhero whose mighty body can deflect a volley of bullets fired from a machine gun by a villain. As they strike and bounce off of our superhero, these bullets would exert force on the body. The microscopic collisions of individual gas particles on the containing walls similarly exert force, macroscopically manifested as pressure. In the following, we will develop this idea mathematically and derive the ideal gas law.

When a particle bounces off the wall, its direction changes and its velocity goes from $-v$ to v (or from v to $-v$). This change in velocity, $\Delta v = v - (-v) = 2v$, comes from the force acting on the particle by the wall. The particle, in turn, exerts a force with the same magnitude and opposite direction on the wall, according to Newton's Third Law of Motion.

According to Newton's Second Law of Motion, this force F is equal to the mass of the particle m times its acceleration (the rate of change in its velocity) a, so $F = ma$. Acceleration of an object is the rate of change of velocity, or $\frac{\Delta v}{\Delta t}$, and we had also discussed that $\Delta t = 2L/|v|$. Therefore,

$$F_{\text{single particle}} = m\frac{\Delta v}{\Delta t} = m\frac{2v|v|}{2L}.$$

The magnitude of the force is $|F_{\text{single particle}}| = mv^2/L$.

Since there are many particles with different velocities that would collide with the wall, we should consider the average force delivered by N different particles. Let's use a notation of a pair of angled brackets, $< \cdot >$ to denote an average quantity. Then, for an average total force on the wall due to N particles:

$$< F >= N\frac{m < v^2 >}{L}.$$

By dividing this average force by the cross-sectional area A of the wall, we can find pressure $P = F/A$. We also note $V = LA$, where V is the volume of the container, so

$$P = \frac{< F >}{A} = N\frac{m < v^2 >}{LA} = Nm\frac{< v^2 >}{V}.$$

We can rewrite the above expression as:

$$PV = Nm < v^2 > .$$

Our discussion so far has been for one-dimensional motion, so v is really v_x, a horizontal component of the total, three-dimensional velocity $\vec{v}$. Since $|\vec{v}|^2 = v_x^2 + v_y^2 + v_z^2$ and since x-dimension is not any more special than y- or z-dimension, the average of horizontal velocity squared $< v^2 >$ which we have been discussing so far is equal to one-third of the average total velocity squared $\frac{1}{3} < |\vec{v}|^2 >$.

Our expression is then

$$PV = \frac{1}{3}Nm < |\vec{v}|^2 > = N\frac{2}{3} < \text{Kinetic Energy} >,$$

where we made an identification that the kinetic energy of a particle is given by $\frac{1}{2}mv^2$.

By comparing this relationship with the ideal gas law $PV = NkT$, we arrive at a conclusion that

$$< \text{Kinetic Energy} > = \frac{3}{2}kT,$$

and

$$\sqrt{< |\vec{v}|^2 >} = v_{\text{rms}} = \sqrt{\frac{3kT}{m}}.$$

In other words, temperature T is a measure of the average kinetic energy of N gas particles, whose root-mean-square (rms) speed is $\sqrt{3kT/m}$. This is the main result of the Kinetic Theory of Gas.

2.3 SAMPLE CALCULATION

Consider 1 mole of nitrogen gas N_2, which is the most abundant type of gas molecule in the air. Each N_2 has the mass of 4.68×10^{-26} kg. What is the average root-mean-square speed at a temperature of 300 K (approximately room temperature)? The value of the Boltzmann constant k is 1.38×10^{-23} kg m^2 / sec^2 K.

Answer: 515 m/sec. There will be other particles moving faster and slower than this speed.

As a hypothetical situation, if all N_2 particles were moving at this speed and if they were striking a wall of area 100 cm^2 in a head-on collision (i.e., one-dimensional motion) for 0.5 sec, what would be the pressure exerted on the wall by these particles?

Answer: These particles would all experience the change of velocity of 2×515 m/sec and the acceleration of $a = \Delta v / \Delta t = 2060$ m/sec^2. Each particle would exert a force of ma, and since there are N_A molecules, the pressure would be equal to 5.80×10^3 Pa.

2.4 FURTHER EXPLORATIONS

The following block of code compares the average $< x >$ and root-mean-square $x_{\mathrm{rms}} = \sqrt{< x^2 >}$. If x is a collection of N numbers $(x_1, x_2, ..., x_N)$,

$$< x >= \frac{x_1 + x_2 + ... + x_N}{N},$$

and

$$x_{\mathrm{rms}} = \sqrt{\frac{x_1^2 + x_2^2 + ... + x_N^2}{N}}.$$

```
# Code Block 2.4

# How to calculate <x> and x_rms with numpy array.
import numpy as np

x = np.array([0,1,2,3,4])

# You can try 5 random numbers with the following line.
#x = np.random.randn(5)

N = len(x)
print("Number of items = %d"%N)
print("x    = ",np.array2string(x, precision=3, sign=' '))
print("x^2 = ",np.array2string(x**2, precision=3, sign=' '))
print("<x>    = %4.3f"%(np.sum(x)/N))
print("<x^2> = %4.3f"%(np.sum(x**2)/N))
print("x_rms = %4.3f"%(np.sqrt(np.sum(x**2)/N)))
```

```
Number of items = 5
x    =  [0 1 2 3 4]
x^2 =  [ 0   1   4   9 16]
<x>    = 2.000
<x^2> = 6.000
x_rms = 2.449
```

Let's calculate the force exerted on the wall in two ways, using our mechanical interpretation of elastic collisions. One way to calculate the force is to add up the momentum changes, $m\Delta v$, due to multiple collisions by multiple particles on the wall and then to divide the sum by the time Δt. Another way is to use one of the results derived earlier, $<F>= N\frac{m<v^2>}{L}$. We can directly calculate $<v^2>$ from a list of particle velocities at a fixed time.

In other words, the first method requires observing and measuring the collision events at one place (i.e., the wall) for an extended period. On the other hand, the second method requires collecting the velocity values of all gas particles everywhere at one point in time (i.e., a snapshot of velocity distribution). Both methods yield consistent values.

The following code block calculates the force acting on a wall by **N** particles with random velocity distribution. **F1** is based on the momentum change, and **F2** is based on the average value of the velocity squared. To compare these two different calculations, the percent difference between **F1** and **F2** is also calculated.

```
# Code Block 2.5

N = 100 # number of particles

ymin = 0
ymax = 2

tmin = 0
tmax = 10
dt = 0.1

t = np.arange(tmin,tmax,dt)
Nbounce = np.zeros(N)

y0 = np.random.rand(N)
v = np.random.randn(N)*2
v = np.sort(v)
for i in range(N):
    _, _, Nbounce[i] = calculate_position(y0[i],v[i],dt=dt,
                                           ymin=ymin,ymax=ymax,
```

```
                                        tmin=tmin,tmax=tmax)

m = 1 # mass of a particle
L = ymax-ymin
delta_t = tmax - tmin
delta_v = 2*np.abs(v)
v_rms = np.sqrt(np.sum(v**2)/N)

F1 = m*np.sum( Nbounce * delta_v / delta_t )*0.5
# Factor of 0.5, since Nbounce counts the collision with two walls.
print("Force = m(delta v/delta t) = %4.3f"%F1)

F2 = N*m*v_rms**2/L
print("Force = Nm<v^2>/L = %4.3f"%F2)
# Calculate percent difference as (A-B) / average(A,B) * 100.
print("Percent Difference = %3.2f"%(100*(F2-F1)*2/(F1+F2)))

# misc. info.
print('\n===========================================\n')
print('misc. info:')
print('speed range: min = %4.3f, max = %4.3f'
    %((np.min(np.abs(v))),np.max(np.abs(v)))))
print('number of particles with 0 bounce = %d'%(np.sum(Nbounce==0)))
print('number of particles with 1 bounce = %d'%(np.sum(Nbounce==1)))
```

```
Force = m(delta v/delta t) = 184.189
Force = Nm<v^2>/L = 185.549
Percent Difference = 0.74


===============================================

misc. info:
speed range: min = 0.007, max = 6.223
number of particles with 0 bounce = 4
number of particles with 1 bounce = 14
```

It is worthwhile to investigate why the percent difference between two force calculations is not zero. One potential reason is the resolution of the numerical simulation. For example, if **dt** is too big, the temporal resolution of our simulation is too coarse. A fast particle might move between the walls too quickly during a single timestep (that is, $v \cdot dt > L$), and we would not be able to track the particle's position precisely enough and might miscount the collision event. We can decrease **dt** for more temporal resolution, which will make the simulation longer.

Another reason for the mismatch is the discrete nature of a collision. The force is exerted on the wall at the instance of a collision, but the force is zero while the particle moves through the empty space between the walls. During the simulation, some slow particles may not even strike a wall once. Then, these particles would not contribute to the value of force calculated with the number of collisions. At the same time, their slow velocities would still contribute to the value of force calculated with v_{rms}. For other slow particles, if they bounced off from a wall once at the beginning of the simulation but still are very far away from the other wall, they would collide with a wall only once. Their contribution to an average force value would be more than what it would be if the simulation ran for a long time (with a significant value of `tmax`). Such a tradeoff (accuracy versus temporal or spatial resolution) is an important and commonly encountered design consideration for many types of numerical simulations.

2.5 TEMPERATURE

Our derivation of the ideal gas law has revealed that the temperature is directly proportional to the average kinetic energy of gas particles, which is directly proportional to v_{rms}^2.

In the following block of code, we make a comparison between T and v_{rms}^2. For T, we will refer to the ideal gas law, $PV = NkT$, so that $kT = PV/N$. The quantity P will be calculated by measuring the force due to individual collisions of N gas particles in the simulation, as we have done above. Then, $PV = (F/A)(AL) = FL$, where A is the area of the wall and L is the distance between the two walls.

The root-mean-square velocity, v_{rms}, can simply be calculated by `np.sqrt(np.sum(v**2)/N)`, since in this simulation, each particle maintains its speed. The particles only collide with the walls, so their velocities will always be either v or $-v$. In the next chapter, we will consider what happens when particles collide.

We will simulate different temperature values by generating random velocities with different magnitudes, which is accomplished by `v = np.random.randn(N)*T`, so that high T increases `v_rms`. In other words, higher temperature results in particles traveling faster with more collision events, which in turn would be manifested as increased pressure on the wall.

The following simulation verifies that the kinetic theory of gas "works" over different situations. The simulation parameter T scales the range of velocities by v = np.random.randn(N)*T, so high T increases v_rms. As the resulting plot demonstrates, PV/N is indeed equal to $m < v^2 >$ or mv_{rms}^2. The data points lie along a diagonal line. The factor of 3 in the derivation of an ideal gas is not here because our simulation is one-dimensional.

```python
# Code Block 2.6

def run_with_different_T (T=1,N=100,m=1):
    ymin = 0
    ymax = 2
    L = ymax-ymin # distance between the walls.

    tmin=0
    tmax=10
    dt = 0.1

    t = np.arange(tmin,tmax,dt)
    Nbounce = np.zeros(N)
    v = np.random.randn(N)*T # T scales the v_rms.
    y0 = np.random.rand(N)
    v = np.sort(v)
    for i in range(N):
        _, _, Nbounce[i] = calculate_position(y0[i],v[i],dt=dt,
                                     ymin=ymin,ymax=ymax,
                                     tmin=tmin,tmax=tmax)
    delta_t = tmax - tmin
    delta_v = 2*np.abs(v)
    v_rms = np.sqrt(np.sum(v**2)/N)
    F = m*np.sum( Nbounce * delta_v / delta_t )*0.5
    PV = F*L
    return PV, v_rms

T_range = np.arange(0.1,5,0.2) # Range of temperatures to consider.
N = 100 # number of particles.
m = 1 # mass of each particle.

perc_diff = np.zeros(len(T_range))
maxval = 0 # keep track of max value for scaling the plot.
for i,T in enumerate(T_range):
    PV, v_rms = run_with_different_T(T=T,N=N,m=m)
    plt.scatter(m*v_rms**2,PV/N,color='black')
    # Calculate percent difference as (A-B)/average(A,B)*100.
    perc_diff[i] = (m*v_rms**2 - PV/N)*2/(m*v_rms**2+PV/N)*100
    if maxval < PV/N:
        maxval = PV/N
```

```
# Draw a diagonal line.
plt.plot([0,maxval*1.1],[0,maxval*1.1],color='black',linewidth=1)
plt.xlabel('m*v_rms^2')
plt.ylabel('PV/N')
plt.axis('square')
plt.savefig('fig_ch2_KineticTheorySummary.eps')
plt.show()

print("Mean Percent Difference = %3.2f"%np.mean(perc_diff))
```

```
Mean Percent Difference = 1.49
```

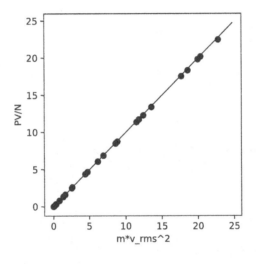

Figure 2.6

In conclusion, the kinetic theory of gas successfully explains how macroscopic properties of an ideal gas (P and T) arise out of microscopic events. Gas pressure comes from numerous elastic collisions between the container and the gas molecules. The temperature of a gas is a manifestation of the average kinetic energy of gas particles.

CHAPTER 3

Velocity Distribution

3.1 PARTICLE COLLISION

A movement of a particle can be described by its velocity, $\vec{v}$, which is a vector quantity with magnitude (how fast it is moving) and direction (where it is heading). The velocity $\vec{v}$ and mass m of a particle determine the particle's momentum ($\vec{p} = m\vec{v}$) and kinetic energy ($mv^2/2$). Note v without an arrow at the top means the magnitude of a vector, $|\vec{v}|$. When two particles run into each other during an elastic collision, their momentum and kinetic energy may change, but their combined total momentum and kinetic energy are conserved.

$$m_1\vec{v}_{1,\text{ before}} + m_2\vec{v}_{2,\text{ before}} = m_1\vec{v}_{1,\text{ after}} + m_2\vec{v}_{2,\text{ after}}$$

$$\frac{1}{2}m_1v_{1,\text{ before}}^2 + \frac{1}{2}m_2v_{2,\text{ before}}^2 = \frac{1}{2}m_1v_{1,\text{ after}}^2 + \frac{1}{2}m_2v_{2,\text{ after}}^2$$

Since $\vec{v}$ has x, y, and z components, v_x, v_y, and v_z, the above equations can be written out this way, too.

$$m_1v_{1x,\text{ before}} + m_2v_{2x,\text{ before}} = m_1v_{1x,\text{ after}} + m_2v_{2x,\text{ after}}$$

$$m_1v_{1y,\text{ before}} + m_2v_{2y,\text{ before}} = m_1v_{1y,\text{ after}} + m_2v_{2y,\text{ after}}$$

$$m_1v_{1z,\text{ before}} + m_2v_{2z,\text{ before}} = m_1v_{1z,\text{ after}} + m_2v_{2z,\text{ after}}$$

$$\frac{1}{2}m_1(v^2_{1x,\text{ before}} + v^2_{1y,\text{ before}} + v^2_{1z,\text{ before}})$$
$$+ \frac{1}{2}m_2(v^2_{2x,\text{ before}} + v^2_{2y,\text{ before}} + v^2_{2z,\text{ before}})$$
$$= \frac{1}{2}m_1(v^2_{1x,\text{ after}} + v^2_{1y,\text{ after}} + v^2_{1z,\text{ after}})$$
$$+ \frac{1}{2}m_2(v^2_{2x,\text{ after}} + v^2_{2y,\text{ after}} + v^2_{2z,\text{ after}})$$

There are a lot of possible solutions that simultaneously satisfy the above set of equations because there are 12 variables (three spatial dimensions and two particles for before and after conditions) that are constrained by only four relationships (three for momentum conservation and one for energy conservation).

3.2 ONE-DIMENSIONAL EXAMPLE

Let's consider a head-on (one-dimensional) collision of two particles. Since we assume no movements along other dimensions (that is, $v_y = v_z = 0$ for before and after the collision), we can combine just two equations that describe the conservation of momentum and kinetic energy. After a few lines of algebra, we can solve for the post-collision velocities along the x dimension.

$$v_{1x,\text{ after}} = \frac{m_1 - m_2}{m_1 + m_2}v_{1x,\text{ before}} + \frac{2m_2}{m_1 + m_2}v_{2x,\text{ before}}$$

$$v_{2x,\text{ after}} = \frac{2m_1}{m_1 + m_2}v_{1x,\text{ before}} + \frac{m_2 - m_1}{m_1 + m_2}v_{2x,\text{ before}}$$

The above result is symmetric. When we swap the indices 1 and 2, the resulting expressions remain mathematically identical, as they should be since there is nothing special about which particle is called 1 or 2.

Another fun consideration is to swap "before" and "after" distinctions, and then solve for the post-collision velocities. After a few lines of algebra, we again obtain mathematically identical expressions, as they should be since particle collisions are symmetric under time-reversal. That means that the time-reversed process of a particle collision is possible. If two particles with velocities of $v_{1x,\text{ after}}$ and $v_{2x,\text{ after}}$ collided with each other, their post-collision velocities would be $v_{1x, \text{ before}}$ and $v_{2x, \text{ before}}$. Note it is not a typo that I am calling the post-collision velocities with "before" labels, as we are considering a time-reversed collision event. If we recorded a movie of microscopic particle collisions and played it backward in time, it would look as physically plausible as the original movie, as the conservation laws of momentum and kinetic energy would hold. (Another way to think about the time reversal is to replace t with $-t$ in an equation of motion and to notice that this extra negative sign does not change the equation.)

The profound question is, then, why is there an arrow or directionality of time that we experience macroscopically. For example, a concentrated blob of gas particles would diffuse across the space, as a scent from a perfume bottle would spread across a room, not the other way. This question will be addressed in later chapters as we further develop a statistical description of various physical phenomena.

As an interesting case of a two-particle collision event, let's consider $m_2 \gg m_1$ and $v_{2x, \text{ before}} = 0$. Then, we would have a reasonable result of $v_{1x, \text{ after}} = -v_{1x, \text{ before}}$ and $v_{2x, \text{ after}} = 0$. In other words, m_1 would bounce off of a more massive particle, m_2. Another interesting case is $m_1 = m_2$ and $v_{2x, \text{ before}} = 0$. We obtain $v_{1x, \text{ after}} = 0$ and $v_{2x, \text{ after}} = v_{1x, \text{ before}}$. In other words, the first particle comes to a stop, and the second particle is kicked out with the same speed as the first particle after the collision. We sometimes see this type of collision on a billiard table when a white cue ball hits a ball at rest, gives all its momentum to that ball, and comes to an immediate stop. The last example is $m_1 = m_2$ and $v_{1x, \text{ before}} = -v_{2x, \text{ before}}$, where two particles move toward each other with the same momentum. This collision sends each particle in the opposite direction with the same speed.

The following code draws cartoons of these three examples programmatically. The calculation of post-collision velocities is implemented in the function **headon_collision()**. The plotting routine is packaged in the function **plot_pre_post_collision()**, which splits a figure window into two subplots. The left subplot draws the pre-collision condition, and the right subplot shows the post-collision condition. Each particle is placed with a **scatter()** command, and its velocity is drawn with an **arrow()** command. A **for**-loop is used to issue the same commands for constructing each subplot.

The following code also illustrates the use of the **assert** command, which can be used to check test cases and potentially catch programming errors.

```
# Code Block 3.1

import matplotlib.pyplot as plt

def headon_collision (u1,u2,m1,m2):
    # u1 and u2: pre-collision velocities
    # v1 and v2: post-collision velocities
    v1 = ((m1-m2)*u1 + 2*m2*u2)/(m1+m2)
    v2 = (2*m1*u1 + (m2-m1)*u2)/(m1+m2)
    return v1, v2

def plot_pre_post_collision(velocity1,velocity2,m1,m2):
    x1, x2 = (-1,1) # Location on m1 and m2.
    y = 1 # Arbitrary location along y.
    marker_size = 100
    scale = 0.5 # the length scale of the velocity arrow.
    title_str = ('Before','After')
    c1 = 'black'
    c2 = 'gray'
    fig,ax = plt.subplots(1,2,figsize=(8,1))
    for i in range(2):
        ax[i].scatter(x1,y,s=marker_size,color=c1)
        ax[i].scatter(x2,y,s=marker_size,color=c2)
        # Draw an arrow if the velocity is not too small.
        if abs(velocity1[i])>0.01:
            ax[i].arrow(x1,y,velocity1[i]*scale,0,color=c1,
                        head_width=0.1)
        if abs(velocity2[i])>0.01:
            ax[i].arrow(x2,y,velocity2[i]*scale,0,color=c2,
                        head_width=0.1)
        ax[i].set_xlim((-2,2))
        ax[i].set_ylim((0.8,1.2))
        ax[i].set_title(title_str[i])
        ax[i].axis('off')
    plt.tight_layout()

print('Case 1: m2 is much more massive than m1 and is at rest.')
m1, m2 = (1, 2000) # Like an electron and a proton.
u1, u2 = (1, 0)
v1, v2 = headon_collision(u1,u2,m1,m2)
plot_pre_post_collision((u1,v1),(u2,v2),m1,m2)
plt.savefig('fig_ch3_collision_case1.eps')

print('Case 2: m1 = m2, and m2 is initially at rest.')
```

```
m1, m2 = (1, 1)
u1, u2 = (1, 0)
v1, v2 = headon_collision(u1,u2,m1,m2)
plot_pre_post_collision((u1,v1),(u2,v2),m1,m2)
plt.savefig('fig_ch3_collision_case2.eps')
assert v1==0
assert v2==u1

print('Case 3: m1 = m2, and they move toward each other.')
m1, m2 = (1, 1)
u1, u2 = (1, -1)
v1, v2 = headon_collision(u1,u2,m1,m2)
plot_pre_post_collision((u1,v1),(u2,v2),m1,m2)
plt.savefig('fig_ch3_collision_case3.eps')
assert v1==-u1
assert v2==-u2
```

Case 1: m2 is much more massive than m1 and is at rest.

Case 2: m1 = m2, and m2 is initially at rest.

Case 3: m1 = m2, and they move toward each other.

Figure 3.1

3.3 MULTIPLE SOLUTIONS

Unlike the simple, one-dimensional example with a single solution, a complete three-dimensional collision has more than one solution. Consider two identical particles with the same mass $m = 1$. Without loss of generality, we may assume that the first particle is moving along the x direction with a velocity of 1, and the second particle is at rest.

Let's use the following notation to keep track of various velocity components:

$$(\vec{v}_1, \vec{v}_2)_{\text{before}} = (v_{1x}, v_{1y}, v_{1z}, v_{2x}, v_{2y}, v_{2z})_{\text{before}} = (1, 0, 0, 0, 0, 0).$$

Now consider the following set of velocities:

$$(v_{1x}, v_{1y}, v_{1z}, v_{2x}, v_{2y}, v_{2z})_{\text{after}} = (0.9, 0.3, 0.0, 0.1, -0.3, 0.0).$$

We can verify that momentum along each of the three dimensions is conserved, because

$$v_{1x, \text{ after}} + v_{2x, \text{ after}} = 0.9 + 0.1 = 1,$$
$$v_{1y, \text{ after}} + v_{2y, \text{ after}} = 0.3 - 0.3 = 0,$$
$$v_{1z, \text{ after}} + v_{2z, \text{ after}} = 0.$$

Furthermore, the total kinetic energy after the collision is

$$\frac{1}{2}m(v_{1x}^2 + v_{1y}^2 + v_{1z}^2 + v_{2x}^2 + v_{2y}^2 + v_{2z}^2)$$
$$= \frac{1}{2}(0.9^2 + 0.3^2 + 0.1^2 + 0.3^2)$$
$$= \frac{1}{2}(0.81 + 0.09 + 0.01 + 0.09)$$
$$= \frac{1}{2}(1.0^2),$$

which is the same as the initial kinetic energy of $\frac{1}{2}(1.0^2)$.

There are other sets of numbers that satisfy the same constraints: $(0.8, 0.4, 0.0, 0.2, -0.4, 0.0)$, $(0.5, 0.5, 0.0, 0.5, -0.5, 0.0)$, $(0.5, 0.0, -0.5, 0.5, 0.0, 0.5)$, and others.

We have demonstrated that there are multiple solutions. Here we will further show that the total kinetic energy of the system can be shared between the two particles in various proportions.

As before, we will use a set of six numbers to denote the velocity components of two particles, $(v_{1x}, v_{1y}, v_{1z}, v_{2x}, v_{2y}, v_{2z})$, and assume that the pre-collision velocity is $(1, 0, 0, 0, 0, 0)$. The following is a list of post-collision velocities and the ratio of kinetic energies, $E_1 : E_2$, which is equal to $\frac{1}{2}m(v_{1x}^2 + v_{1y}^2 + v_{1z}^2) : \frac{1}{2}m(v_{2x}^2 + v_{2y}^2 + v_{2z}^2)$.

$(v_{1x}, v_{1y}, v_{1z}, v_{2x}, v_{2y}, v_{2z})$ after collision	$E_1 : E_2$
(+1.0, +0.0, +0.0, +0.0, +0.0, +0.0)	1.0:0.0
(+0.9, +0.3, +0.0, +0.1, -0.3, +0.0)	0.9:0.1
(+0.8, +0.4, +0.0, +0.2, -0.4, +0.0)	0.8:0.2
(+0.5, +0.5, +0.0, +0.5, -0.5, +0.0)	0.5:0.5
(+0.5, +0.3, +0.4, +0.5, -0.3, -0.4)	0.5:0.5
(+0.2, +0.4, +0.0, +0.8, -0.4, +0.0)	0.2:0.8
(+0.1, +0.3, +0.0, +0.9, -0.3, +0.0)	0.1:0.9
(+0.0, +0.0, +0.0, +1.0, +0.0, +0.0)	0.0:0.1
(+0.1, +0.3, +0.0, +0.9, -0.3, +0.0)	0.1:0.9
(+0.0, +0.0, +0.0, +1.0, +0.0, +0.0)	0.0:0.1

The above outcomes are possible post-collision velocities, as they satisfy both momentum and energy conservation principles. Different solutions have different values for x, y, and z velocity components, indicating that these particles are moving toward or away from each other at different angles. These solutions also show that depending on how these two particles approach and recoil at different angles, the total momentum and energy may be split up in many different ways after the collision. The first particle may end up with most of the energy, and it is also possible for the second particle to leave with more energy.

3.4   FINDING SOLUTIONS WITH CODE

Over the following several blocks of code, we will systematically and exhaustively search for possible velocities. Even though brute-force search may not be the most mathematically sophisticated or elegant, it can be relied on when an analytical method is not readily available.

We will develop a few versions of the search code so that the subsequent version will be faster than the previous ones. The main idea will be the same. We will use six **for**-loops to go through possible values of $(v_{1x}, v_{1y}, v_{1z}, v_{2x}, v_{2y}, v_{2z})_{\text{after}}$, and check whether they jointly satisfy the principles of momentum and energy conservation. The improvements of different code versions will come from reducing the search space. As an analogy, suppose you are trying to find a key you lost in the house, and you can optimize your search by only visiting the places you've been to recently. That is, don't look for your key in the bathroom if you haven't been there.

There are two main variables in the following code: **v_before** and **v_after**, each of which contains six numbers, and they are put into individual variables to make the code more readable. For example, **v1x_b** corresponds to $v_{1x, \text{ before}}$ and **v2z_a** corresponds to $v_{2z, \text{ after}}$. Each of these six numbers in **v_before** and **v_after** can be accessed using an index value between 0 and 5 (0, 1, 2, 3, 4, 5). Multiple values can be accessed with the : symbol. For example, **v_before[:3]** returns the first three values from the full list: **v_before[0]**, **v_before[1]**, and **v_before[2]**, starting with index 0 and ending before index 3. These three values are **v1x_b**, **v1y_b**, and **v1z_b**. Similarly, **v_before[3:]** returns the remaining three values, starting with index 3, which are assigned to **v2x_b**, **v2y_b**, and **v2z_b**.

We can check momentum conservation by asking if the difference between the sum of pre-collision velocities and the sum of post-collision velocities is zero. However, numerical values do not have infinite precision. For example, we know numbers like $\sqrt{2}$, π, and 1/3 are infinitely long in their decimal representations, such as 1.414..., 3.141592..., and 0.333..., but a typical numerical computation deals with a fixed number of decimal points, possibly producing small but non-zero round-off errors. (Note that there are methods that deal with infinite-precision or arbitrary-precision arithmetic, which we are not using.) Therefore, in our implementation, we will check whether the absolute value of the difference is smaller than a relatively tiny number **tol**, rather than comparing the difference to zero.

The following code block illustrates this idea: **if (num1-num2)==0:** checks whether **num1** and **num2** are numerically identical, and **if abs(num1-num2)<tol:** checks whether these two values are close enough up to a tolerance threshold value **tol**.

```
# Code Block 3.2

# Illustration of comparing two numbers,
# while considering limited numerical precision.

num1 = 1.0/3.0
num2 = 0.333333333
tol = 0.0001

print('Compare two numbers')
print('num1 = ',num1)
print('num2 = ',num2)

if (num1-num2)==0:
    print('num1 is equal to num2.')
else:
    if abs(num1-num2)<tol:
        print('num1 is practically equal to num2.')
    else:
        print('num1 is not equal to num2.')
```

```
Compare two numbers
num1 =  0.3333333333333333
num2 =  0.333333333
num1 is practically equal to num2.
```

In the following function `is_conserved()`, if `np.abs( (v1x_b+v2x_b) - (v1x_a+v2x_a) ) < tol:` checks whether the momentum along x-direction is conserved within a numerical tolerance threshold. Note that we are assuming the mass of all particles is the same, so we can just compare the velocities, instead of the product of mass and velocity. Similarly, `if np.abs(e_b - e_a) < tol:` checks energy conservation. If all four conditions (3 momentum conservations and 1 energy conservation) are passed, `v_after` is a solution, so the `is_conserved()` function returns `True`, and if any one of the conservation checks fails, the function returns `False`.

```
# Code Block 3.3

import numpy as np

def is_conserved(v_before,v_after,tol=0.0001):
    v1x_b, v1y_b, v1z_b = v_before[:3]
    v2x_b, v2y_b, v2z_b = v_before[3:]
    v1x_a, v1y_a, v1z_a = v_after[:3]
    v2x_a, v2y_a, v2z_a = v_after[3:]
```

```
# defining kinetic energy
e_b = v1x_b**2+v1y_b**2+v1z_b**2+v2x_b**2+v2y_b**2+v2z_b**2
e_a = v1x_a**2+v1y_a**2+v1z_a**2+v2x_a**2+v2y_a**2+v2z_a**2
if np.abs( (v1x_b+v2x_b) - (v1x_a+v2x_a) ) < tol:
    if np.abs( (v1y_b+v2y_b) - (v1y_a+v2y_a) )< tol:
        if np.abs( (v1z_b+v2z_b) - (v1z_a+v2z_a) ) < tol:
            if np.abs(e_b - e_a) < tol:
                return True
    return False

# A few simple test cases.
v_before = np.array([1,0,0,0,0,0])
assert is_conserved(v_before,np.array([0,0,0,1,0,0]))==True
assert is_conserved(v_before,np.array([0,0,0,1.1,0,0]))==False
assert is_conserved(v_before,np.array([0.5,0.5,0,0.5,-0.5,0]))==True
assert is_conserved(v_before,np.array([0.5,0.5,0,0.5,0.5,0]))==False
```

The following block has a function `generate_solutions_very_slow()`, which takes `v_before` as an input and considers a range of possible values between -`maxval` and `maxval` for six different velocity values with six nested **for**-loops. These six values are packaged into the `v_after` variable and passed to the `is_conserved()` function, which checks whether the momentum and energy are conserved. If `v_after` passes the checks, it is appended onto a list named `solutions`, and then the next set of numbers is considered.

The range of values to be considered is created by `np.arange(-maxval,maxval+dv,dv)`. The maximum value `maxval` is given by the square root of the sum of squares of all pre-collision velocities: `np.sqrt(np.sum(np.array(v_before)**2))` or $\sqrt{|\vec{v}_{1,\,before}|^2 + |\vec{v}_{2,\,before}|^2}$, because if any of the post-collision velocity was bigger than this value, the energy conservation principle would not hold. For example, suppose `v_before` was $(1,0,0,0,0,0)$ and hence, `maxval` was equal to 1. Then if any one of `v_after` was greater than 1 (let's say, $(1.1,0,0,0,0,0)$), the post-collision kinetic energy would be greater than the pre-collision kinetic energy. Therefore, the solution would be found only between -`maxval` and `maxval`.

Another important parameter in this function is `dv` which dictates how thoroughly we will look for a solution within the range. For example, suppose `maxval = 1` and `dv = 0.5`. Then, for each velocity component, we would try five different values: -1, -0.5, 0, 0.5, and 1. For `dv = 0.1`, we would try 21 values: -1, -0.9, -0.8, ..., 0.8, 0.9, and 1. For `dv = 0.01`, there are 201 values. The number of values to consider is proportional to $1/dv$. Because there are six nested **for**-loops, the total number of possible combinations for **v_after** goes up exponentially like $(1/dv)^6$.

This exponential behavior makes **generate_solutions_very_slow()** very slow and inefficient. For **maxval = 1** and **dv = 0.5**, this function will process more than 15625 $(=5^6)$ cases. For **dv = 0.1**, the number of cases would be more than 85 million, and for **dv = 0.01**, the number would increase to more than 6.5×10^{13}. In fact, do not use this function unless you can leave your computer running for many, many hours.

```
# Code Block 3.4

def generate_solutions_very_slow (v_before,dv=0.1):
    maxval = np.sqrt(np.sum(np.array(v_before)**2))
    num = 0
    solutions = list()
    for v1x in np.arange(-maxval,maxval+dv,dv):
        for v1y in np.arange(-maxval,maxval+dv,dv):
            for v1z in np.arange(-maxval,maxval+dv,dv):
                for v2x in np.arange(-maxval,maxval+dv,dv):
                    for v2y in np.arange(-maxval,maxval+dv,dv):
                        for v2z in np.arange(-maxval,maxval+dv,dv):
                            v_after = [v1x,v1y,v1z,v2x,v2y,v2z]
                            # keep track of number of trials.
                            num = num+1
                            if is_conserved(v_before,v_after):
                                solutions.append(v_after)
    return solutions, num

# You can make dv smaller,
# but the code will take a really long time.
dv = 0.5
solutions, num = generate_solutions_very_slow([1,0,0,0,0,0],dv=dv)
print("Number of solutions tried = ", num, ' with dv = ', dv)
assert num==(5**6)
```

Number of solutions tried = 15625 with dv = 0.5

The following function **generate_solutions()** has a similar structure of the six nested **for**-loops (to consider possible sets of post-collision velocities) and one **if** statement (to check conservation). The difference with the previous implementation is that it narrows down the search space further, so that the code runs significantly faster. Because the total energy needs to be conserved, the inner **for**-loops consider the range of velocity values that the remaining energy can accommodate. As an analogy, suppose we have a set of siblings, and they are given a total of \$100. The oldest sibling (A) takes a portion of it (say, \$40) and gives the rest to the next oldest one (B), who takes a portion from the remaining fund and passes the rest to the next one (C). We know that A+B+C=$100, so B cannot have more than $60 because $40 was already taken by A.

A helper function `my_range()` generates a reduced range for each nested loop. In the first `for`-loop for `v1x`, we consider the full range between `-maxval` and `maxval`, but in the second loop for `v1y`, we consider a smaller range since `v1x` reduced the range of possible values for `v1y`. The next loop for `v1z` considers an even smaller range since both `v1x` and `v1y` took up a portion of energy, and a smaller amount of energy is left for `v1z`. Therefore, the input argument for `my_range()` starts with `[0,0,0,0,0,0]` and incrementally adds the contributions from the previous loops: `[v1x,0,0,0,0,0]`, `[v1x,v1y,0,0,0,0]`, `[v1x,v1y,v1z,0,0,0]`, etc.

The reduced range is given by `np.arange(-new_maxval,new_maxval+dv,dv)`, where `new_maxval` is the remaining amount of energy. That is the square root of `maxval**2-np.sum(np.array(trial)**2)`, where `trial` is the set of velocity values that are being considered in the previous, outer `for`-loops. In the above sibling analogy, this `trial` would be a record of how much money was already taken by the older siblings.

There are a few lines of code that may need some clarification. The `tmp_maxval` is a temporary variable for the square root of `maxval**2-np.sum(np.array(trial)**2)`, which, due to rare unfortunate numerical round-off errors, may be slightly less than 0, so `my_sqrt()` function returns zero in such a case.

Why might we encounter such a case? That is because the code was written so that any trial solutions we consider would come from a range of `np.arange(-maxval,maxval+dv,dv)`, a set of evenly-spaced numbers. If the newly returned `tmp_maxval` happens to be a number not contained in this discrete set, we pick the closest number using the `np.argmin()` function to determine `new_maxval`, thereby introducing a small numerical imprecision.

Compared to `generate_solutions_very_slow()`, this new function `generate_solutions()` is a significant improvement. For `maxval = 1` and `dv = 0.1`, the slow code considers more than 85 million possibilities, while this one considers 5,827,203 possibilities, which is a speed-up of more than ten-fold. A geometric insight of this optimization is this. The slow code searches for a solution within the full volume of a (six-dimensional) cube, while the new code searches on a surface of a (six-dimensional) sphere.

```
# Code Block 3.5

def my_range(maxval,trial,dv):

    def my_sqrt (v):
        return np.sqrt(np.max([v,0]))

    def_range = np.arange(0,maxval+dv,dv) # Default range
    tmp_maxval = my_sqrt(maxval**2-np.sum(np.array(trial)**2))
    new_maxval = def_range[np.argmin(np.abs(def_range-tmp_maxval))]
    return np.arange(-new_maxval,new_maxval+dv,dv)

def generate_solutions (v_before,dv=0.1):
    maxval = np.sqrt(np.sum(np.array(v_before)**2))
    num = 0
    solutions = list()
    trial = [0,0,0,0,0,0]
    for v1x in my_range(maxval,trial,dv):
        trial = [v1x,0,0,0,0,0]
        for v1y in my_range(maxval,trial,dv):
            trial = [v1x,v1y,0,0,0,0]
            for v1z in my_range(maxval,trial,dv):
                trial = [v1x,v1y,v1z,0,0,0]
                for v2x in my_range(maxval,trial,dv):
                    trial = [v1x,v1y,v1z,v2x,0,0]
                    for v2y in my_range(maxval,trial,dv):
                        trial = [v1x,v1y,v1z,v2x,v2y,0]
                        for v2z in my_range(maxval,trial,dv):
                            trial = [v1x,v1y,v1z,v2x,v2y,v2z]
                            v_after = np.array(trial)
                            num = num+1
                            if is_conserved(v_before,v_after):
                                solutions.append(v_after)
    return solutions, num

dv = 0.5
solutions, num = generate_solutions([1,0,0,0,0,0],dv=dv)
print("Number of solutions tried = ", num, ' with dv = ', dv)
```

```
Number of solutions tried =  545  with dv =  0.5
```

The use of **generate_solutions()** is illustrated here. A helper function, **print_solutions_neatly()**, takes the solutions list and prints them out with a fixed-width format. The **datetime** module is used to keep track of the runtime.

```
# Code Block 3.6

from datetime import datetime

def print_solutions_neatly (solutions):
    for sol in np.array(solutions):
        print(np.array2string(sol, precision=2, suppress_small=True,
                              floatmode='fixed', sign=' '))
    return

# Pre-collision velocities.
v_before = [1,0,0,0,0,0] # [v1x,v1y,v1z,v2x,v2y,v2z]

# Set the resolution of search.
# Recommended dv = 0.1.
dv = 0.1
# Calculation takes significantly longer (hours) with small dv.
#dv = 0.05

print("\nTrying out generate_solutions() function.")
print("Started at ====> ", datetime.now().strftime("%H:%M:%S"))
solutions, num = generate_solutions (v_before,dv=dv)
print("Finished at ===> ", datetime.now().strftime("%H:%M:%S"))
print_solutions_neatly(solutions)
print("Number of solutions tried = ", num, ' with dv = ', dv)
```

```
Trying out generate_solutions() function.
Started at ====>   10:25:45
Finished at ===>   10:26:34
[-0.00 -0.00 -0.00  1.00 -0.00 -0.00]
[ 0.10 -0.30 -0.00  0.90  0.30 -0.00]
[ 0.10 -0.00 -0.30  0.90  0.00  0.30]
[ 0.10 -0.00  0.30  0.90  0.00 -0.30]
[ 0.10  0.30 -0.00  0.90 -0.30 -0.00]
[ 0.20 -0.40 -0.00  0.80  0.40 -0.00]
[ 0.20 -0.00 -0.40  0.80 -0.00  0.40]
[ 0.20 -0.00  0.40  0.80 -0.00 -0.40]
[ 0.20  0.40 -0.00  0.80 -0.40 -0.00]
[ 0.50 -0.50 -0.00  0.50  0.50 -0.00]
[ 0.50 -0.40 -0.30  0.50  0.40  0.30]
[ 0.50 -0.40  0.30  0.50  0.40 -0.30]
[ 0.50 -0.30 -0.40  0.50  0.30  0.40]
[ 0.50 -0.30  0.40  0.50  0.30 -0.40]
```

```
[ 0.50 -0.00 -0.50  0.50 -0.00  0.50]
[ 0.50 -0.00  0.50  0.50 -0.00 -0.50]
[ 0.50  0.30 -0.40  0.50 -0.30  0.40]
[ 0.50  0.30  0.40  0.50 -0.30 -0.40]
[ 0.50  0.40 -0.30  0.50 -0.40  0.30]
[ 0.50  0.40  0.30  0.50 -0.40 -0.30]
[ 0.50  0.50 -0.00  0.50 -0.50 -0.00]
[ 0.80 -0.40 -0.00  0.20  0.40 -0.00]
[ 0.80 -0.00 -0.40  0.20 -0.00  0.40]
[ 0.80 -0.00  0.40  0.20 -0.00 -0.40]
[ 0.80  0.40 -0.00  0.20 -0.40 -0.00]
[ 0.90 -0.30  0.00  0.10  0.30 -0.00]
[ 0.90 -0.00 -0.30  0.10  0.00  0.30]
[ 0.90 -0.00  0.30  0.10  0.00 -0.30]
[ 0.90  0.30  0.00  0.10 -0.30 -0.00]
[ 1.00 -0.00 -0.00 -0.00 -0.00 -0.00]
Number of solutions tried =  5827203  with dv =  0.1
```

```
# Code Block 3.7

# Slower function.
# Uncomment, if you want to try this out.
v_before = [1,0,0,0,0,0] # [v1x,v1y,v1z,v2x,v2y,v2z]
dv = 0.1
#print("\nTrying out generate_solutions_very_slow() function.")
#print("Started at ====> ", datetime.now().strftime("%H:%M:%S"))
#solutions, num = generate_solutions_very_slow (v_before,dv=dv)
#print("Finished at ===> ", datetime.now().strftime("%H:%M:%S"))
#print_solutions_neatly(solutions)
#print("Number of solutions tried = ", num, ' with dv = ', dv)
```

3.5 DISTRIBUTION OF ENERGY

The above calculation shows many possible solutions to a two-particle collision problem. The post-collision velocities may take on various values; hence, the two particles can split the total energy in many different ways after the collision. If a shopper goes into a store with some amount of cash, the shopper may spend different amounts of money on other days. The total cash in the possession of the shopper and the store should be equal to the starting amount of the shopper, but this entire amount may be split in many different ways.

The following code plots the split of the total energy as a histogram. It calculates the energy ratio as the amount of post-collision kinetic energy carried away by one particle, divided by the total pre-collision kinetic energy,

$$\frac{E_{1,\ \text{after}}}{E_{1,\ \text{before}} + E_{2,\ \text{before}}}.$$

The histogram shows that this ratio can be 0 (if all the post-collision kinetic energy is given to the second particle), 0.5 (if the energy is shared equally), 1 (if the first particle takes all the kinetic energy after the collision), or other values.

```python
# Code Block 3.8

# Draw a histogram of energy ratio.
energy_ratio = np.zeros((len(solutions),2))
for i, sol in enumerate(solutions):
    e_before = np.sum(np.array(v_before)**2)
    e1_after = np.sum(np.array(sol[:3])**2)
    e2_after = np.sum(np.array(sol[3:])**2)
    energy_ratio[i,0] = e1_after/e_before
    energy_ratio[i,1] = e2_after/e_before

# Make a histogram for the first particle with
# the values in the first column of energy_ratio array.
plt.hist(energy_ratio[:,0],bins=11)
plt.title('Histogram of energy ratio')
plt.xlabel('$E_1$ (after) / Total Energy')
plt.ylabel('Number')
plt.xticks((0,0.5,1))
plt.savefig('fig_ch3_energy_ratio_hist.eps')
plt.show()
```

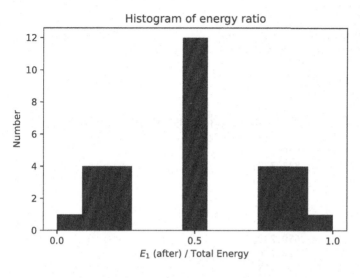

Figure 3.2

3.6 DISTRIBUTION OF ENERGY AFTER MANY, MANY COLLISION EVENTS

Then, an interesting question is if there is a stable (i.e., equilibrium) distribution of energy of many particles. In other words, even if only one particle was moving initially, it will collide with other particles, and its kinetic energy will start to spread and be shared with others. Over time, the velocity distribution profile will fluctuate because a fast particle may get slowed down and a slow particle may gain extra energy from a random collision with a particle with higher kinetic energy. After a sufficiently long time, it turns out that this distribution takes on a stable shape.

In the following simulation, we will generate this distribution by randomly splitting energies, based on the earlier observation that energy splits in many different ways. As a preliminary step, let's work on a code that picks two particles randomly. We will use **np.random.randint()**, which picks random integer values within a specified range. For example, **np.random.randint(5,size=(10,2))** will create ten pairs of random integers between 0 and 5.

```
# Code Block 3.9

N = 1000 # Number of particles.
Tmax = 10 # Number of iterations (or number of picks)
random_picks = np.random.randint(N,size=(Tmax,2))
for t in range(Tmax):
    pick1 = random_picks[t,0]
    pick2 = random_picks[t,1]
    print("Random pick %2d = (%3d,%3d)"%(t,pick1,pick2))
```

```
Random pick  0 = (909,322)
Random pick  1 = (827,778)
Random pick  2 = (818,165)
Random pick  3 = (145,948)
Random pick  4 = (536,847)
Random pick  5 = (950, 65)
Random pick  6 = (609,213)
Random pick  7 = (123,756)
Random pick  8 = (293,440)
Random pick  9 = (170,308)
```

We will start the simulation where two particles are chosen randomly, exchanging their energies while keeping the total energy constant. What would be the resulting distribution of energy? In this simulation, we pick two particles `i` and `j` using the `np.random.randint()` function. If these particles are different (`if (i!=j):`), split their total energy (`e[i]+e[j]`) randomly. This random split is accomplished by picking a number between 0 and 1 (`p = np.random.rand()`). This is a simplifying assumption because the histogram of energy split from our earlier calculation is not uniform. Still, it simplifies our simulation into a single line of code.

```
# Code Block 3.10

import numpy as np
import matplotlib.pyplot as plt

# Simulation of how particles exchange their energy over time.

N = 1000 # number of particles
T = 2000 # number of iterations.

e = np.zeros(N) # Current energy distribution of N particles.
e_all = np.zeros((T,N)) # Energy distribution over time.

# Initialize all particles with the same energy.
e0 = 10
```

```
e = e + e0

# Here is another way to initialize the energy distribution:
# Let's give just one particle all energy.
#e = np.zeros(N)
#e[0] = e0*N


for t in range(T):
    e_all[t,:] = e

    # Pick a pair of particles.
    i = np.random.randint(N)
    j = np.random.randint(N)
    if (i!=j):
        # Exchange energy if two different particles are chosen.
        e_total = e[i]+e[j]
        p = np.random.rand() # fraction (between 0 and 1)
        e[i] = np.round(e_total*p) # Give a portion of energy to i.
        e[j] = e_total-e[i] # Give the rest of energy to j.
```

```
# Code Block 3.11

# Show histogram at the beginning and end.
def show_hist (e_all, t):
    T,N = e_all.shape
    width = 5
    maxlim = np.ceil(np.max(e_all)*width)/width
    bins = np.arange(0,maxlim,width)
    h, b = np.histogram(e_all[t,:], bins=bins)
    plt.bar(b[:-1],h/N,width=width,color='black')
    plt.ylim((0,1.1))
    plt.ylabel('Fraction of Particles')
    plt.yticks((0,0.5,1))
    plt.xlabel('Energy (a.u.)')
    plt.title('Distribution of energy')

print('Distribution at the beginning')
show_hist(e_all,0)
plt.savefig('fig_ch3_energy_dist_initial.eps')
plt.show()

print('Distribution after a long time (equilibrium)')
show_hist(e_all,-1)
plt.savefig('fig_ch3_energy_dist_final.eps')
plt.show()
```

Distribution at the beginning

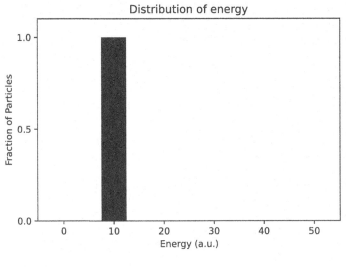

Figure 3.3

Distribution after a long time (equilibrium)

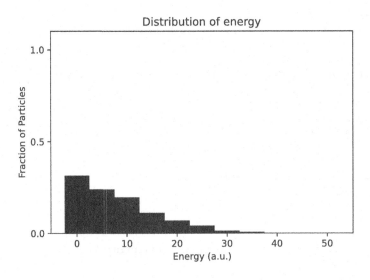

Figure 3.4

```
# Code Block 3.12
# Show histogram over time.

import numpy as np
import matplotlib.pyplot as plt

T,N = e_all.shape
width = 5
maxlim = np.ceil(np.max(e_all)*width)/width
bins = np.arange(0,maxlim,width)

nrow = 5
ncol = 5
fig, axes = plt.subplots(nrow,ncol,figsize=(8,8),
                         sharex=True,sharey=True)
step = int(T/(nrow*ncol))
for i in range(nrow):
    for j in range(ncol):
        ax = axes[i,j]
        t = (i*ncol + j)*step
        h, b = np.histogram(e_all[t,:], bins=bins)
        ax.bar(b[:-1],h/N,width=width,color='black')
        ax.set_title("t = %d"%t)
        ax.set_yticklabels([])
        ax.set_xticklabels([])
        ax.set_ylim((0,1.1))

# Put axes labels on the last subplot.
ax = axes[i,0]
ax.set_yticks((0,0.5,1))
ax.set_ylabel('Fraction of Particles')
ax.set_xlabel('Energy (a.u.)')
plt.tight_layout()
plt.savefig('fig_ch3_energy_dist_evolution.eps')
plt.show()
```

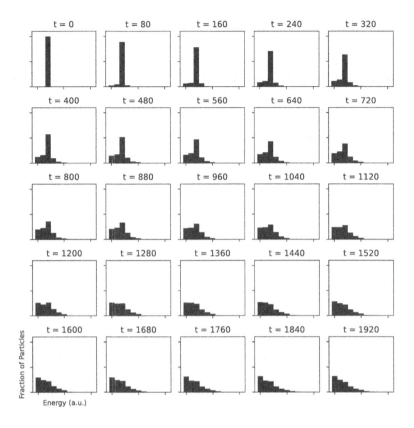

Figure 3.5

Figure 3.5 summarizes the results. After many energy exchanges through random elastic collisions, the gas particles share and distribute the total energy in a particular way. The stable distribution of energy ϵ, as shown by the histogram, looks like an exponential function, $e^{-\beta\epsilon}$. (Note we will use ϵ to denote a continuous variable for energy.) According to the shape of this distribution, most particles have small energy, but there are chances for some particles to carry high kinetic energies.

We will write the probability of a particle having kinetic energy between ϵ and $\epsilon + d\epsilon$ as $P(\epsilon)d\epsilon$. $P(\epsilon)$ is called the probability density function. The energy distribution can be written as $P(\epsilon)d\epsilon \propto e^{-\beta\epsilon}d\epsilon$, where the proportionality constant is determined by the normalization condition

$1 = \int_0^\infty P(\epsilon)d\epsilon$. This normalization comes from the fact that the energy of any particle will certainly be between 0 and ∞.

The exponential behavior of $P(\epsilon)$ can also be inferred by the following observations. Consider a collision of two particles with pre- and post-collision energies, $(\epsilon_{1, \text{ before}}, \epsilon_{2, \text{ before}})$ and $(\epsilon_{1, \text{ after}}, \epsilon_{2, \text{ after}})$. The probability of a collision is proportional to the product of individual probabilities. Furthermore, the inverse collision process, where the pre- and post-collision energy values are swapped, should also be equally probable at equilibrium. Hence, we have:

$$P(\epsilon_{1, \text{ before}})P(\epsilon_{2, \text{ before}}) = P(\epsilon_{1, \text{ after}})P(\epsilon_{2, \text{ after}}),$$

or

$$\ln P(\epsilon_{1, \text{ before}}) + \ln P(\epsilon_{2, \text{ before}}) = \ln P(\epsilon_{1, \text{ after}}) + \ln P(\epsilon_{2, \text{ after}}).$$

We also require energy conservation.

$$\epsilon_{1, \text{ before}} + \epsilon_{2, \text{ before}} = \epsilon_{1, \text{ after}} + \epsilon_{2, \text{ after}}.$$

By the inspection of the last two expressions, we can infer that $P(\epsilon) \propto e^{-\beta\epsilon}$. We will come back to this result with more rigorous proof later.

3.7 DISTRIBUTION OF SPEED AFTER MANY, MANY COLLISION EVENTS

We will assume a bit of mathematical sophistication in the following analysis. We will use the previous result of exponential energy distribution and derive the distribution of the speed of gas particles.

Velocity is a vector in three-dimensional space, and its magnitude, $|\vec{v}|$ (or v), is speed. Therefore, a distribution of velocities involves a three-dimensional probability density, while the energy distribution obtained above is one-dimensional.

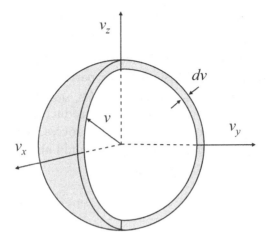

Figure 3.6

Assuming isotropy (that is, all directions are equal or there is no special direction), we may consider a thin spherical shell in the velocity phase space (where the axes are v_x, v_y, and v_z). A spherical shell whose radius is $v = \sqrt{v_x^2 + v_y^2 + v_z^2}$ and whose thickness is dv will have a differential volume of $4\pi v^2 dv$, and the velocities represented within this volume will have the same energy. Then, the distribution of speed is given by:

$$P(v)dv \propto 4\pi v^2 e^{-\beta \frac{mv^2}{2}} dv.$$

This result is called the Maxwell-Boltzmann distribution. The exponential factor of this distribution comes from $e^{-\beta \epsilon}$ of energy distribution, because the energy and speed are related according to $\epsilon = \frac{1}{2}mv^2$. The factor, $4\pi v^2 dv$, arises from the spherical geometry. The volume of a spherical shell with thickness dv is greater for a large v. That means that the possible range of v is higher for a large v for a given interval dv. For example, there are more ways of obtaining the speed between 100 and $100 + dv$ (large v) than between 50 and $50 + dv$ (smaller v). More ways correspond to a higher probability.

This idea is illustrated in Figure 3.7, which shows two circular bands drawn with the same thickness. The area of the circular band with a larger radius is larger than the one with a smaller radius.

```
# Code Block 3.13

import matplotlib.pyplot as plt

circle1 = plt.Circle((0,0), 2.0, alpha=1, color='gray')
circle2 = plt.Circle((0,0), 1.5, alpha=1, color='white')
circle3 = plt.Circle((0,0), 1.0, alpha=1, color='black')
circle4 = plt.Circle((0,0), 0.5, alpha=1, color='white')

fig, ax = plt.subplots()
ax.add_patch(circle1)
ax.add_patch(circle2)
ax.add_patch(circle3)
ax.add_patch(circle4)

plt.axis('equal')
plt.axis('off')
plt.savefig('fig_ch3_more_area.eps')
plt.show()
print('The gray band covers more area than the black band, ')
print('even though they have the same thickness.')
```

The gray band covers more area than the black band,
even though they have the same thickness.

Figure 3.7

```
# Code Block 3.14

def MB_distribution (v,T=1,dv=0.01):
    # Define probability distribution function.
    # Parameter in the exponential distribution
    # is given by dimensionless temperature T.
    p = v**2*np.exp(-v**2/T)
    p = p/(np.sum(p)*dv) # normalize.
    return p

dv = 0.01
v = np.arange(0,5,dv)
plt.plot(v,MB_distribution(v,T=1,dv=dv),color='k',linestyle='solid')
plt.plot(v,MB_distribution(v,T=2,dv=dv),color='k',linestyle='dotted')
plt.plot(v,MB_distribution(v,T=4,dv=dv),color='k',linestyle='dashed')
legend_txt = ('Cold (or Heavy)','Warm (or Medium)','Hot (or Light)')
plt.legend(legend_txt,framealpha=1)
plt.xlabel('v (a.u.)')
plt.ylabel('P(v)')
plt.yticks((0,0.5,1))
plt.title('Maxwell-Boltzmann Distribution')
plt.savefig('fig_ch3_maxwell_boltzmann_distrib.eps')
plt.show()
```

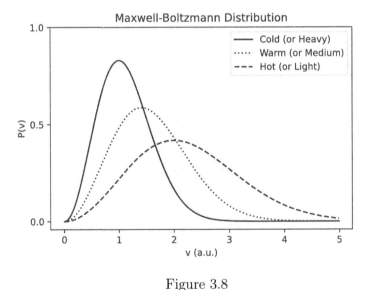

Figure 3.8

Figure 3.8 shows the Maxwell-Boltzmann distribution. The rising and falling behavior comes from two opposing trends: the disfavoring of large

v due to the exponential distribution of energy and the favoring of large v due to the three-dimensional spherical nature of v.

Furthermore, when the parameter $1/\beta$ is large, the distribution shifts and stretches to the right. This situation corresponds to a higher temperature of the gas or a larger amount of total kinetic energy. The proportionality constant can also be derived from the normalization condition of the probability density, $1 = \int_0^\infty P(v)dv$. Because the total area under the curve should be 1, the distribution curve shrinks vertically as it stretches horizontally.

A similar effect of curve-shifting happens with different species of the ideal gas with different mass m at the same temperature. According to the kinetic theory of gas, the average kinetic energy is the same for all gases with the same temperature. However, lighter gas particles would move faster than heavier gas with the same amount of kinetic energy. Thus, a lighter gas such as helium would have a speed distribution that is shifted and stretched toward larger v, and a heavier gas would be shifted toward smaller v.

3.8   NOTE ON A MORE AMBITIOUS CODING PROJECT

It is possible to simulate the collision of particles more realistically. Such a coding project may involve the following steps.

1. N particles are randomly distributed in a volume, V (or a two-dimensional area, for easier visualization with a scatter plot). They are initialized with random velocities.

2. At each time step, update the position of each particle with two important considerations. First, if a particle runs into a wall (i.e., the new position goes outside of the boundary of V), assume that an elastic collision with an immovable wall has occurred and the particle bounces back into V. Second, if two particles come too close to each other (i.e., $|\vec{r}_1 - \vec{r}_2| < 2r$, where $|\vec{r}_1 - \vec{r}_2|$ is the distance between two particles and r is the radius of each particle), assume an elastic collision has occurred and update their positions and velocities by applying energy and momentum conservation principles.

3. Run the simulation forward in time, and compile the velocity distributions, which converge to a Maxwell-Boltzmann distribution.

However, constantly updating the positions of N particles and checking possible collisions of $N(N-1)/2$ pairs of particles are computationally intensive, so a thoughtful search optimization also needs to be implemented.[†]

[†]There are many excellent tutorials and examples of such simulation of particle dynamics. For a reference, see:

- "Building Collision Simulations" by Reducible:
  www.youtube.com/watch?v=eED4bSkYCB8
- A collision simulation can be tweaked to model an epidemic (as done by 3B1B):
  www.youtube.com/watch?v=gxAa02rsdIs
- Python code example: "The Maxwell–Boltzmann distribution in two dimensions" at scipython.com/blog

CHAPTER **4**

Thermal Processes

4.1 STATE AND PROCESS

An earlier chapter examined a model of ideal gas as a collection of N particles flying around in random directions and colliding with one another. This kinetic theory of gas is a satisfyingly intuitive model as such motions and collisions are familiar to us from a game of billiards, bowling, or curling. At any given point in time, the state of an ideal gas can be specified with its current values of P (pressure), V (volume), and T (temperature). This chapter will consider processes where these state variables change smoothly. Such changes may come from adding or subtracting heat energy and applying force (e.g., compression on the container) on the gas.

Among many possible thermal processes, a few simple ones are particularly important and amenable to theoretical analysis. An isobaric process deals with a change in V and T while P stays constant (i.e., by putting the gas in a flexible container that would expand or contract easily, so that it would be at equilibrium with a constant external pressure). The prefix "iso" means "equal," and "bar" is an old unit of pressure. A device that measures pressure is called a "barometer." Hence "isobaric" means "of equal pressure."

An isothermal process refers to any change in V and P while T stays constant (i.e., by putting the gas in thermal contact with a big reservoir whose temperature does not change). An isochoric process refers to any change in T and P while V is constant (i.e., by putting the gas in a rigid box that does not expand or contract). An adiabatic process refers to a situation where no heat energy is added or removed from the gas (i.e., by isolating the gas in a heat-insulating container).

The thermal processes of an ideal gas are often visualized as curves on a PV-diagram (a graph whose axes are P and V). The following simulation from PhET is a great resource for thinking about different thermal processes of an ideal gas: `https://phet.colorado.edu/en/simulation/gas-properties`

4.2 PLOTTING AND NUMERICAL INTEGRATION

Before we get to the physics of thermal processes, let's get acquainted with a few plotting techniques. Study and play with the following code, so that you can understand what each line does. An important lesson is that it is straightforward to overlay different types of graphs (a line graph and a bar graph in this case) with the `matplotlib.pyplot` module.

The resulting graph provides a visual illustration of how the area under a function, or an integral, can be approximated by the sum of the areas of many rectangles with small widths. This numerical integration yields a better approximation with the rectangles of smaller widths, but at the expense of having to deal with more terms in the calculation and, hence, a longer runtime.

```python
# Code Block 4.1

import numpy as np
import matplotlib.pyplot as plt

dx = 5
x = np.arange(0,100+dx,dx)
y = x**2/100

# Other functions you may try.
# y = np.log(0.01*x+0.1)+3
# y = 2*x + 150

# Calculate the area under curve
# by adding up the areas of the small rectangles.
A = np.sum(y)*dx
print('area under curve:', A)

plt.plot(x,y,'o-',color='k')
```

```
plt.bar(x,y,width=dx*0.85,color='gray')
# width = how wide each bar is.

plt.ylim([-dx, np.max(y)+dx])
plt.xlim([-dx, np.max(x)+dx])
plt.xlabel('x')
plt.ylabel('y')
plt.savefig('fig_ch4_numerical_integ.eps')
plt.show()
```

```
area under curve: 3587.5
```

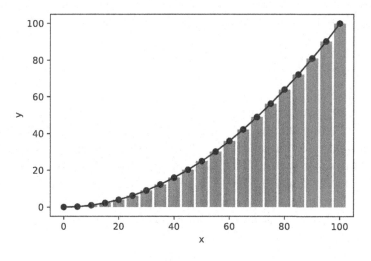

Figure 4.1

4.3 PV DIAGRAM

The study of thermal processes is highly relevant to the engineering of thermal engines like an internal combustion engine, where the expansion of gas produces useful work (such as the rotation of the wheels of a car) at the expense of thermal energy. It is also convenient to imagine gas trapped inside a piston that can expand in one direction with a fixed cross-sectional area A. The force exerted by the gas is PA, and if the gas moves the piston by dx, the volume change is $dV = Adx$. According to the definition of mechanical work, the total amount of work performed by the expanding gas can be calculated by

$$W = \int \vec{F} \cdot d\vec{x} = \int_{x_a}^{x_b} (PA)dx = \int_{V_a}^{V_b} PdV.$$

The pressure and volume of gas can change in many different ways. One of the simplest ways is an isobaric process where the gas expands while maintaining its pressure. For example, a piston may be allowed to push out slowly against the constant atmospheric pressure. This is analogous to a case where you lift a weight by "fighting against" the constant gravitational force, mg, over distance h. The amount of work you did would be mgh. The total amount of work performed by the gas during an isobaric process, where the volume changes from V_a to V_b, is:

$$W_{\text{isobaric}} = \int_{V_a}^{V_b} PdV = P(V_b - V_a).$$

Another case is an isochoric process, where V does not change even when P or T are changing. You may want to visualize this situation as a gas trapped inside a rigid box with immovable walls. Then $dV = 0$, so the total work is zero. This is analogous to a situation where you are trying to lift a weight, but it is so heavy that you cannot move it, no matter how much force you exert. Despite your assertion of force, no mechanical work was done because the displacement of the weight is zero.

$$W_{\text{isochoric}} = \int_{V_a}^{V_a} PdV = 0.$$

Another case is an isothermal process where the pressure and volume are inversely proportional, as seen from the ideal gas law in Chapter 2.

$$W_{\text{isothermal}} = \int_{V_a}^{V_b} PdV = \int_{V_a}^{V_b} \frac{NkT}{V} dV = NkT \ln \frac{V_b}{V_a}.$$

4.4 ADIABATIC PROCESS

An adiabatic process is a bit more complicated. During an isothermal expansion, the gas maintains its temperature, which means that extra thermal energy is added to the gas at the same rate as the gas does the work, so that the average internal energy of the gas (i.e., its temperature) is maintained. However, during an adiabatic expansion, such infusion of energy is not allowed by definition. This would be analogous to a case where you lift a weight without any caloric intake. Over the long run, you are depleting your internal energy and will not be able to do as much work. In other words, compared to the isothermal expansion process over the same range of volume change, an adiabatic expansion process will produce less work and end up at a lower temperature.

As a consequence, ideal gas that goes through an adiabatic process between states a and b has an additional constraint, in addition to the usual $PV = NkT$:

$$PV^\gamma = \text{constant, or } P_a V_a^\gamma = P_b V_b^\gamma,$$

where $\gamma = C_P/C_V$ is a ratio of specific heat coefficients under constant pressure and constant volume. It is $5/3$ for an ideal monoatomic gas. The proof of this result is a bit involved, but it touches upon many interesting insights, so here it is.

4.5 PROOF OF PV^γ = CONSTANT FOR AN ADIABAT OF IDEAL GAS

Let's start with a very general statement that a thermal system can be specified with its state variables, T, V, and P. If these variables are related by an equation of state variables (for example, $PV = NkT$ for an ideal gas), the system is ultimately specified with only two variables. Considering the internal energy $U(T, V)$ as a multivariate function of T and V, we can make the following general statement:

$$dU = \left(\frac{\partial U}{\partial T}\right)_V dT + \left(\frac{\partial U}{\partial V}\right)_T dV.$$

The expression $\left(\frac{\partial U}{\partial T}\right)_V$ is a partial derivative of U with respect to T while keeping V constant. Likewise, $\left(\frac{\partial U}{\partial V}\right)_T$ is a partial derivative of U with respect to V under constant T.

Another general statement we can make is that

$$dU = dQ + dW = dQ - PdV.$$

The above statement means that the internal energy of a thermal system would change from the heat transfer (dQ) or the mechanical work done on the system (dW). The latter is $dW = Fdx = -PdV$, as shown in the chapter on the kinetic theory of gas. The negative sign of $-PdV$ indicates that if the gas is compressed (negative dV), work is done on the system by the external force, and hence dU should be positive.

How universal are those two statements? Do T and V completely specify the internal energy of a physical system? Is there any other way (beyond dQ and dW) to change the internal energy of a physical system? There could be other relevant state variables and other ways to affect the internal energy. For example, the internal energy of some material may be affected by the absorption or emission of a photon. Other material may change internal energy when it is subject to electrical potential. However, these are sufficiently rigorous statements within our context of classical thermodynamics.

Now let's focus our context even more narrowly by considering ideal gas. According to the kinetic theory of gas, the internal energy of ideal gas is the sum of an individual particle's kinetic energy, which determines T, so U of ideal gas would not depend on V. Therefore, $\left(\frac{\partial U}{\partial V}\right)_T = 0$. Furthermore, because $PV = NkT$, $PdV + VdP = NkdT$, which will be used shortly.

Now, using the first two expressions about dU, we have:

$$dQ - PdV = \left(\frac{\partial U}{\partial T}\right)_V dT + \left(\frac{\partial U}{\partial V}\right)_T dV.$$

$$dQ = PdV + \left(\frac{\partial U}{\partial T}\right)_V dT.$$

$$dQ = (-VdP + NkdT) + \left(\frac{\partial U}{\partial T}\right)_V dT = -VdP + \left[Nk + \left(\frac{\partial U}{\partial T}\right)_V\right] dT.$$

It is convenient to define specific heat, $C_V = \left(\frac{dQ}{dT}\right)_V$, which is the amount of heat energy needed to raise the temperature of a system at constant volume. Similarly, $C_P = \left(\frac{dQ}{dT}\right)_P$ is defined as the specific heat at constant pressure. Given the above relationships, we see that for an ideal gas, $C_V = \left(\frac{\partial U}{\partial T}\right)_V$ and $C_P = C_V + Nk$. This latter relationship is called Mayer's equation and indicates that $C_P > C_V$. When the thermal system is allowed to expand (that is, V is not constant), the added heat energy will not only go into the internal energy of the system but also will be spent through its mechanical work of pushing against the environment. Therefore, a system allowed to expand will need more heat energy to increase its temperature, compared to a different system whose volume does not change.

Now we have

$$dQ = C_V dT + P dV,$$

and

$$dQ = C_P dT - V dP.$$

During an adiabatic process, $dQ = 0$, so $P dV = -C_V dT$ and $V dP = C_P dT$. By combining these two expressions,

$$\gamma \equiv \frac{C_P}{C_V} = \frac{-V dP}{P dV}.$$

$$\frac{dP}{P} + \gamma \frac{dV}{V} = 0.$$

By integrating the above variable-separated expression,

$$\ln P + \ln V^\gamma = \ln(PV^\gamma) = \text{constant.}$$

In other words, we have shown that PV^γ is constant for ideal gas during an adiabatic process.

Finally, since the total internal energy of ideal gas is $\frac{3}{2}NkT$ according to the kinetic theory of gas, $C_V = \left(\frac{\partial U}{\partial T}\right)_V = \frac{3}{2}Nk$. Furthermore, according to the Mayer's equation, $C_P = C_V + Nk = \frac{5}{2}Nk$. Therefore, $\gamma = \frac{5}{3}$, as claimed above.

```
# Code Block 4.2

# Draw P-V curves for different thermal processes.
Va = 5
Pa = 20
NkT = Va*Pa

Vb = 20

dV = 0.1
V = np.arange(Va,Vb,dV)

P_isobaric = np.zeros(len(V))+Pa
```

```
P_isotherm = NkT/V
P_adiabat = Pa*(Va/V)**(5/3)

plt.plot(V,P_isobaric,color='black',linestyle='solid')
plt.plot(V,P_isotherm,color='black',linestyle='dotted')
plt.plot(V,P_adiabat,color='black',linestyle='dashed')
plt.legend(('isobaric','isothermal','adiabatic'),framealpha=1.0)
plt.xlim((0,25))
plt.ylim((0,25))
plt.xlabel('V (m$^3$)')
plt.ylabel('P (pascal)')
plt.savefig('fig_ch4_thermal_processes.eps')
plt.show()
```

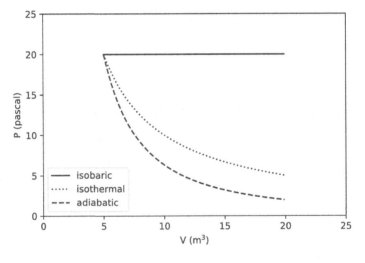

Figure 4.2

4.6 CARNOT CYCLE

A heat engine (or a refrigerator, which is a heat engine running in reverse) can be devised by combining different thermal processes into a cycle. The Carnot cycle is a particularly important example, and it is composed of two sets of alternating isothermal and adiabatic processes.

Imagine ideal gas enclosed in a cylinder with a movable piston. The Carnot cycle starts by warming up the gas with a high-temperature heat source. The added heat increases the gas's internal energy and the gas expands at constant temperature (isothermal expansion). Next, the cylinder with the gas is detached from the heat source, but the gas continues to expand and perform mechanical work. However, since no heat energy is added, the gas temperature drops during this process (adiabatic expansion). During the next portion of the cycle, external work is done on the gas by pushing the piston into the cylinder and thereby compressing the gas. At the same time, the gas temperature is kept at a constant low temperature by bringing the gas-contained cylinder in contact with a low-temperature heat sink and by allowing the heat to exit from the gas (isothermal compression). During the final portion of the cycle, the compression of the gas continues, but the heat sink is removed, and the gas temperature rises without exchanging heat with its environment until the gas returns to its initial thermodynamic state (adiabatic compression). Then, the cycle repeats. The following code block visualizes the Carnot cycle.

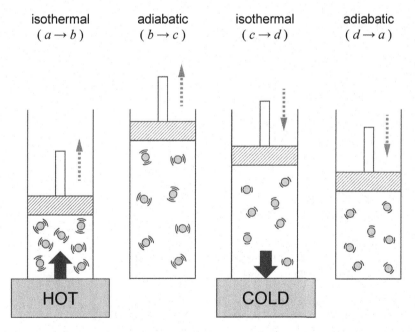

Figure 4.3

```
# Code Block 4.3

# Draw a PV diagram of Carnot cycle,
# going through four different thermal states of a, b, c, and d.

Nk = 1 # for simplicity, let Nk = 1 so that PV = T
gamma = 5/3

dV = 0.1

# Starting points defined below.
Pa = 20  # pressure at the initial state a.

Va = 10  # volume at a.
Vb = 20  # volume at b.
Vc = 40  # volume at c.
Vd = 20  # volume at d.

V_ab = np.arange(Va,Vb,dV)
V_bc = np.arange(Vb,Vc,dV)
V_cd = np.arange(Vc,Vd,-dV)
V_da = np.arange(Vd,Va,-dV)

# Along isotherm (a->b and c->d): P = T/V
# Along adiabat (b->c and d->a): P*V**gamma = constant = k
#       (Note that P*V**gamma = T*V**(gamma-1) since P = T/V)

T_high = Va*Pa # high T
Pa = T_high/Va
P_ab = T_high/V_ab # isothermal process

Pb = T_high/Vb
kb = T_high*Vb**(gamma-1) # constant along adiabat
P_bc = kb/V_bc**(gamma) # adiabatic process

Pc = kb/Vc**(gamma)
T_low = Vc*Pc # low T
P_cd = T_low/V_cd # isothermal process

Pd = T_low/Vd
kd = T_low*Vd**(gamma-1) # constant along adiabat
P_da = kd/V_da**(gamma) # adiabatic process

plt.plot(V_ab,P_ab,color='gray',linestyle='solid')
plt.plot(V_bc,P_bc,color='black',linestyle='dotted')
plt.plot(V_cd,P_cd,color='gray',linestyle='solid')
plt.plot(V_da,P_da,color='black',linestyle='dotted')
plt.legend(('a->b: isothermal','b->c: adiabatic',
            'c->d: isothermal','d->a: adiabatic'),framealpha=1)
```

```
spacing = 1
plt.text(Va+spacing,Pa,'a')
plt.text(Vb+spacing,Pb,'b')
plt.text(Vc+spacing,Pc,'c')
plt.text(Vd+spacing,Pd,'d')
plt.text((Va+Vb)/2+spacing,Pa-6,'high T')
plt.text((Vc+Vd)/2-spacing,Pd-4,'low T')
plt.xlim((0,50))
plt.ylim((0,30))
plt.xlabel('V (m$^3$)')
plt.ylabel('P (pascal)')
plt.savefig('fig_ch4_carnot.eps')
plt.show()
```

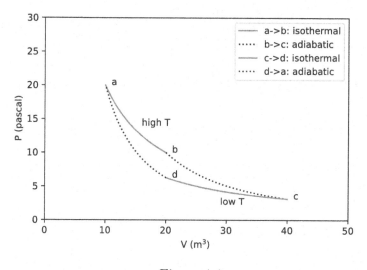

Figure 4.4

As discussed above, mechanical work done during a thermal process can be calculated from $W = \int P dV$, so the total amount of mechanical work done during a full Carnot cycle is equal to the area enclosed by the four curves, composed of two isotherms and two adiabats, in the PV diagram. Because a full cycle brings the gas back to its initial state with the same internal energy, the expended work must have come from the heat energy. The work output would be $W = Q_{in} + Q_{out}$. Here, Q_{in} is the amount of heat that was added to the gas, and Q_{out} is the heat that left the gas. Hence, $Q_{in} > 0$ and $Q_{out} < 0$.

The efficiency of a heat engine or thermal cycle is defined as the ratio of its output (useful mechanical work) and input (total amount of heat energy added during its operation).

$$\eta = \frac{\text{Output}}{\text{Input}} = \frac{W}{Q_{\text{in}}} = \frac{Q_{\text{in}} + Q_{\text{out}}}{Q_{\text{in}}} = 1 + \frac{Q_{\text{out}}}{Q_{\text{in}}}.$$

For a Carnot cycle, the addition of heat energy occurs during the isothermal expansion ($a \rightarrow b$) only, and this thermal energy is equal to the amount of mechanical work performed by the gas as the volume expands from V_a to V_b. These two states are at an equal temperature $T_a = T_b = T_{\text{high}}$.

$$Q_{\text{in}} = \int_{V_a}^{V_b} P dV = NkT_{\text{high}} \ln(V_b/V_a).$$

During the isothermal compression ($c \rightarrow d$), occurring at a constant temperature of $T_c = T_d = T_{\text{low}}$, heat flows out of the gas, and this thermal energy will be equal to

$$Q_{\text{out}} = \int_{V_c}^{V_d} P dV = NkT_{\text{low}} \ln(V_d/V_c).$$

Since $V_d < V_c$, Q_{out} is a negative quantity as expected because the mechanical work is done on the gas when the gas is compressed.

Unlike the two isothermal processes, there is no heat exchange during the adiabatic process ($b \rightarrow c$ and $d \rightarrow a$). Because PV^γ is a constant value along an adiabat, we have $P_b V_b^\gamma = P_c V_c^\gamma$, or $T_b V_b^{\gamma-1} = T_c V_c^{\gamma-1}$ during $b \rightarrow c$. Similarly, $P_d V_d^\gamma = P_a V_a^\gamma$, or $T_d V_d^{\gamma-1} = T_a V_a^{\gamma-1}$ along the path $d \rightarrow a$.

Since $T_a = T_b = T_{\text{high}}$ and $T_c = T_d = T_{\text{low}}$, we can combine the above expressions and obtain $V_b/V_a = V_c/V_d$. Then, $Q_{\text{out}}/Q_{\text{in}} = -T_{\text{low}}/T_{\text{high}}$. Thus, for a Carnot cycle, the efficiency can be calculated just with two operating temperatures:

$$\eta_{\text{Carnot}} = 1 - \frac{T_{\text{low}}}{T_{\text{high}}}.$$

The following code calculates the efficiency of a Carnot cycle in two different ways. The first way is based on the above theoretical formula involving the ratio of temperatures. In the second way, the amount of work done during one cycle is calculated by numerical integration (or the area inside the curve on a PV-diagram), and the amount of added heat is calculated from the amount of mechanical work during isothermal expansion because there is no heat exchange with the environment during the adiabatic processes.

```
# Code Block 4.4
# Calculate the efficiency of a Carnot cycle.

# total amount of work from numerical integration.
W_ab = np.sum(P_ab)*dV
W_bc = np.sum(P_bc)*dV
W_cd = -np.sum(P_cd)*dV
W_da = -np.sum(P_da)*dV
W_total = W_ab+W_bc+W_cd+W_da

# Q_in is equal to the total mechanical work during a->b, because
#    b->c and d->a: adiabatic process, so no added heat.
#    c->d: isothermal compression, so heat goes out, not in.
#    a->b: isothermal expansion, so internal energy does not change.
#          The added heat must match the mechanical work by the gas.
Q_in = W_ab

eta_measure = W_total/Q_in
eta_theory = 1 - T_low/T_high

print("Efficiency of an example Carnot Cycle:")
print("    %4.3f (calculated by numerical integration)"%eta_measure)
print("    %4.3f (according to the theory)"%eta_theory)
print("Percent Difference = %3.2f perc with dV = %4.3f"
      %((1-eta_measure/eta_theory)*100,dV))
print("Smaller dV would make percent difference smaller.")
```

```
Efficiency of an example Carnot Cycle:
    0.381 (calculated by numerical integration)
    0.370 (according to the theory)
Percent Difference = -2.91 perc with dV = 0.100
Smaller dV would make the percent difference smaller.
```

The result obtained with numerical calculation is satisfyingly comparable to the theoretical result. Again, more accurate values can be obtained by improving numerical integration with smaller steps dV.

II

Statistical Mechanics

## CHAPTER 5

Premise of Statistical Mechanics

5.1 ANALOGY: WEALTH DISTRIBUTION

Consider three societies with different wealth distributions, as depicted in the following mock histograms. These hypothetical societies may have the same average wealth per person (i.e., total wealth divided by the total population), but have very different economic profiles. In the first society (a), most of the wealth is concentrated in a small population who are "super rich" (small rectangle) while most people are below the average line (large rectangle). In the second society (b), represented by the rectangles in the middle, an equal fraction of the population is above, below, and at the average wealth level, as shown by three equal-sized rectangles. In the third society (c), most people live with an average income, and only a tiny fraction of the population is above or below the average level. The lifestyle of an average person will be very different in these societies.

Here is a short code sample for creating bar graphs depicting three different distributions.

```
# Code Block 5.1

# Make vertical and horizontal bar histograms of
# hypothetical wealth distribution.

import numpy as np
import matplotlib.pyplot as plt

# Example of different distributions.
dist1 = np.array([70,25,5])
dist2 = np.array([100,100,100])/3
dist3 = np.array([10,80,10])
dists = (dist1,dist2,dist3)

titles = ('(a)','(b)','(c)')
xaxis = ('Low','Med','High')
fig, axes = plt.subplots(1,3,sharey=True,figsize=(8,3))
# Draw a vertical bar plot.
axes[0].bar(xaxis,dist1,color='black')
axes[1].bar(xaxis,dist2,color='black')
axes[2].bar(xaxis,dist3,color='black')
axes[0].set_ylabel('Percent')
for i in range(3):
    axes[i].set_title(titles[i])
    axes[i].set_ylim((0,100))
    axes[i].set_yticks((0,50,100))
    axes[i].set_xlim((-1,3))
plt.tight_layout()
plt.savefig('fig_ch5_distrib_vertical.eps')
plt.show()

xaxis = ('Low','Med','High')
fig, axes = plt.subplots(1,3,sharey=True,figsize=(8,3))
# Draw a horizontal bar plot.
for i in range(3):
    axes[i].barh(xaxis,dists[i],color='black')
    axes[i].set_title(titles[i])
    axes[i].set_xlim((0,100))
    axes[i].set_xticks((0,50,100))
    axes[i].set_ylim((-1,3))
    axes[i].set_xlabel('Percent')
axes[0].set_ylabel('Levels')
plt.tight_layout()
plt.savefig('fig_ch5_distrib_horizontal.eps')
plt.show()
```

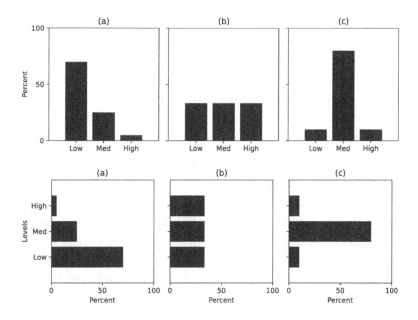

Figure 5.1

In addition to the relative population distributions of income levels, there are other factors. For example, what are the income levels of the rich and the poor in the population? Is the gap between them large or small? How many distinct income brackets should be considered? Is it too simplistic to consider just three (low, medium, and high) levels? If so, how many levels are appropriate to consider? If you are a policy maker, the type of economic policies that you will design and implement will vary greatly, depending on which society you are involved in.

The key point in this analogy is that to have a deep enough understanding of a complex system like societal economics, a single average value (like the average per-capita income) is insufficient. On the other hand, a particular person's wealth is an unnecessarily detailed and impractical piece of information. A good resolution to perform a meaningful analysis is somewhere in the middle, dealing with the brackets of individual wealth and the distribution of the population into these brackets.

Similarly, grade distribution of a class (how many students got high, medium, and low grades on a recent exam) would help a teacher to prepare the lessons for her next class. A CEO of a company would strategize the operation of her company based on a meaningfully informative revenue report with enough, but not too many, details.

Some of the deep understandings of a many-particle system in physics can be drawn with a similar analytical approach, where the knowledge about the available energy levels and the probabilistic distribution of particles in these levels leads to an overall statistical understanding.

5.2 MATHEMATICAL NOTATIONS

Here we will introduce some mathematical notations used throughout the remaining chapters. Just as we discussed different income groups (low, medium, and high), we may consider discrete energy levels with energy ϵ_i. Probability for a particle to occupy this i-th level can be denoted as p_i. N denotes the total number of particles, and the average number of particles in the i-th level would be given by $n_i = Np_i$.

The following conditions hold:

$$1 = \sum_i p_i.$$

$$N = \sum_i n_i.$$

$$U = \sum_i n_i \epsilon_i,$$

$$< U > = \sum_i \epsilon_i p_i,$$

where U is the total energy shared by the particles and $< U >$ is the average energy. N and U are conserved quantities.

When the particles are allowed to exchange their energies with each other freely, they settle into a particular distribution called the Boltzmann distribution. That is, at equilibrium, the number of particles in each energy level takes on a stable mathematical form, as shown here:

$$n_i = \alpha e^{-\epsilon_i/kT},$$

where the normalization condition can determine α, $N = \sum_i n_i$. The meaning of the constants, k and T, will be discussed more later, but for now, they make ϵ_i/kT a unitless quantity. According to this exponential expression, lower energy levels tend to be occupied by more particles, and the higher energy levels have fewer particles.

5.3 LISTING PERMUTATIONS

To develop some intuitions about splitting total energy U by N particles, let's consider a simple example where three people are splitting five \$1 bills. There are several different ways of thinking about all possibilities. One way is to list the amount of money held by each person. If we want to distinguish individuals but not the bills, we may adopt a notation, \$(A, B, C) = \$ amount of money of 3 individuals, so that $A+B+C = \$5$. Here are all possible combinations:

$$(5,0,0),$$
$$(4,1,0),\ (4,0,1),$$
$$(3,2,0),\ (3,1,1),\ (3,0,2),$$
$$(2,3,0),\ (2,2,1),\ (2,1,2),\ (2,0,3),$$
$$(1,4,0),\ (1,3,1),\ (1,2,2),\ (1,1,3),\ (1,0,4),$$
$$(0,5,0),\ (0,4,1),\ (0,3,2),\ (0,2,3),\ (0,1,4),\ (0,0,5).$$

Out of the above list of 21 possibilities, let's count how likely it is for A to have all five bills. Just 1 out of all 21 possible ways. How about four bills? 2 possible ways. How about three? 3. How about two? 4. How about just one? 5. How about no bills? 6. It is more likely for a person to have a smaller amount of money. Their corresponding probabilities would be

$$p(\$5) = 1/21,$$
$$p(\$4) = 2/21,$$
$$p(\$3) = 3/21,$$
$$p(\$2) = 4/21,$$
$$p(\$1) = 5/21,$$
$$p(\$0) = 6/21.$$

Another way to think about such a situation, which we will use more often, is to consider the different possible states of the wealth of each individual. What makes this different from the previous listing is that now we are not distinguishing individuals. We can use the notation of

$(n_0, n_1, n_2, n_3, n_4, n_5)$, where n_i is the number of people having i-number of \$1 bills. More generally, we would call n_i as an occupation number of the i-th state. When these numbers are added together, it should be equal to the total number of people: $n_0 + n_1 + n_2 + n_3 + n_4 + n_5 = 3$. Also, when we count all the \$1 bills distributed among three people, the total should be exactly \$5, equaling the original amount of money. Then, all possibilities of distributing three indistinguishable individuals into these states are:

$$(2,0,0,0,0,1),$$
$$(1,1,0,0,1,0),$$
$$(1,0,1,1,0,0),$$
$$(0,2,0,1,0,0),$$
$$(0,1,2,0,0,0).$$

The first possibility $(2,0,0,0,0,1)$ can be constructed by giving one individual all five bills $(n_5 = 1)$, which leaves no other option than to place the other two individuals with no bills $(n_0 = 2)$. Similarly, the next possibility $(1,1,0,0,1,0)$ is constructed by giving one individual four bills $(n_4 = 1)$ and giving the remaining bill to one individual $(n_1 = 1)$. The remaining one does not get any bills $(n_0 = 1)$. You can systematically tabulate all possibilities in this way. This listing method is similar to the starting example of distributing wealth in a society, and it is again revealed that a smaller amount of money is more likely. In other words, the "occupancy" of states with lower values is higher since it creates more ways of splitting up the fixed total amount.

By the way, we will always assume that the individuals are indistinguishable, as the indistinguishability would be valid for the gas molecules contained in a fixed volume. It is a subtle idea that intrigued many physicists, which was resolved with the advent of quantum mechanical interpretation.

5.4   VISUALIZATION

Here is a sample code for generating a visualization of levels (boxes) and their occupancies (dots). We will draw gray boxes that represent the energy levels with `plt.fill()`, where the width and height of each box are defined as `w` and `h` in the code. In order to keep the boxes separated, a margin value `marg` is subtracted. In each box, we put the specified number of dots with `plt.scatter()`. These dots are scattered

randomly, using **np.random.uniform()**. In the following sample illus-
trations, we place the different number of dots specified in **n**, showing
different occupations at each level. We call each of these distinct configurations microstates.

```
# Code Block 5.2

# Visualize the energy level and occupation numbers.
import numpy as np
import matplotlib.pyplot as plt

def sketch_occupancy (n):
    # Makes a cartoon of occupancy plot.
    # Boxes: levels or states (e.g., number of bills)
    # Dots: occupation numbers (e.g., number of people)

    # Define the size of boxes
    marg = 0.1 # Size of margin
    h = 1.0-2*marg
    w = 1.0-2*marg
    xbox = np.array([marg,marg+w,marg+w,marg])
    ybox = np.array([marg,marg,marg+h,marg+h])

    N = len(n) # number of levels
    for i in range(N):
        plt.fill(xbox,ybox+i,color='#CCCCCC')
        x = (np.random.uniform(size=n[i])-0.5)*w*0.9+0.5
        y = (np.random.uniform(size=n[i])-0.5)*h*0.9+0.5+i
        plt.scatter(x,y,marker='.',color='k',s=50,zorder=2.5)

    plt.ylim(-0.5,N+0.5)
    plt.yticks(ticks=np.arange(N)+0.5,labels=np.arange(N)+1)
    plt.xticks([])
    #plt.ylabel('Energy Levels')
    plt.axis('equal')
    plt.title("Occupancy:\n%s"%n)
    plt.box(on=False)

# Try out different configurations (microstates).

n = [2,0,0,0,0,1]
fig = plt.figure(figsize=(2,8))
sketch_occupancy (n)
plt.savefig('fig_ch5_occupancy_ex1.eps')
plt.show()

n = [1,1,0,0,1,0]
fig = plt.figure(figsize=(2,8))
```

```
sketch_occupancy (n)
plt.savefig('fig_ch5_occupancy_ex2.eps')
plt.show()

n = [0,1,2,0,0,0]
fig = plt.figure(figsize=(2,8))
sketch_occupancy (n)
plt.savefig('fig_ch5_occupancy_ex3.eps')
plt.show()
```

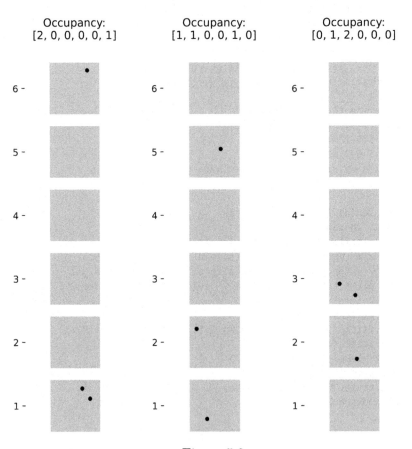

Figure 5.2

5.5 COUNTING EXERCISE

Before we develop a computational routine to list all possible combinations, let's find an upper bound or an overestimate of the number of all possibilities. For simplicity, we can start by assuming that total wealth U can only be split into non-negative integers $(0, 1, 2, \ldots)$ like splitting up money with multiple \$1 bills. We can denote the states of wealth or levels as $\epsilon_0, \epsilon_1, \ldots, \epsilon_{E-1}, \epsilon_E$. There are N indistinguishable people, and they would belong to one of these $E + 1$ states.

To group them, we could line up all N individuals in a single file and insert E dividers at random positions between them. The first divider will mark the boundary between ϵ_0 and ϵ_1, and the next divider will mark the boundary between ϵ_1 and ϵ_2. There are $(N+E)!$ ways of listing N distinguishable individuals and E distinguishable dividers. The first item in a single file could be one of the N individuals or E dividers, so there are $N + E$ possibilities. The next item could be any one of the remaining $(N + E - 1)$ possibilities. Thus, there are $(N + E)(N + E - 1) \cdots (3)(2)(1)$ possible ways of arranging a total of $(N + E)$ distinguishable individuals and dividers. However, because of the indistinguishability of the individuals and dividers, we have to divide out the redundant counts, so the number of combinations is $\frac{(N+E)!}{N!E!}$, which is also called "$N + E$ choose N" or $_{N+E}C_N$ or $\binom{N+E}{N}$. This count is certainly an overestimate because some of these possibilities would not satisfy the constraint that the total wealth is fixed. The following code block illustrates this idea with a few examples.

```
# Code Block 5.3

# Count possible ways of randomly assigning people into groups.

import matplotlib.pyplot as plt
import numpy as np

N = 10 # individuals
E = 5 # dividers, so there will be E+1 categories
# Individuals are represented by zeros.
# Dividers are represented by ones.
arrayN = np.zeros(N)
arrayE = np.ones(E)
# Initial line up of N individuals, followed by E dividers.
x = np.hstack((arrayN,arrayE)).astype('int')
```

```python
Nexamples = 4

print("Randomly place %d individuals into %d groups."%(N,E+1))
print(' ')
for i in range(Nexamples):
    print('Example %d'%(i+1))
    # Now randomly shuffle x.
    x_permute = np.random.permutation(x)
    print(x_permute)
    # Count the number of zeros between ones.
    n = np.zeros(E+1)
    j = 0
    for k in x_permute:
        if k<1: # if not a divider, increment.
            n[j] = n[j]+1
        else:
            print('\tn[%d] = %d'%(j,n[j]))
            j = j+1

    # Take care of the last group.
    print('\tn[%d] = %d'%(j,n[j]))
    print(' ')
```

Randomly place ten individuals into six groups.

```
Example 1
[1 0 0 0 0 0 1 0 0 1 1 1 0 0 0]
        n[0] = 0
        n[1] = 5
        n[2] = 2
        n[3] = 0
        n[4] = 0
        n[5] = 3

Example 2
[0 0 1 0 1 0 0 0 0 0 0 1 1 0 1]
        n[0] = 2
        n[1] = 1
        n[2] = 6
        n[3] = 0
        n[4] = 1
        n[5] = 0

Example 3
[1 1 0 1 0 1 0 0 0 0 0 0 0 1 0]
        n[0] = 0
        n[1] = 0
        n[2] = 1
```

```
        n[3] = 1
        n[4] = 7
        n[5] = 1

Example 4
[0 1 0 1 1 0 0 1 0 0 1 0 0 0 0]
        n[0] = 1
        n[1] = 1
        n[2] = 0
        n[3] = 2
        n[4] = 2
        n[5] = 4
```

5.6 CODE FOR ENUMERATING ALL POSSIBILITIES (VERSION 1)

Now let's develop code for enumerating all possibilities that satisfy our constraints. We will implement two different methods.

The first computational method below is a brute-force algorithm. To systematically go through all possibilities, we consider a $(E + 1)$-digit number in base $(N + 1)$. Each one of the $(E + 1)$ digits can take on a value between 0 and N, and it represents the number of particles in each energy level. Therefore, collectively, these $(E + 1)$ digits correspond to the occupation numbers, $(n_0, n_1, ..., n_{E-1}, n_E)$.

Let's consider a simple example. Suppose there are 9 individual particles $(N = 9)$ that can go into one of three different energy levels: ϵ_0, ϵ_1, and ϵ_2. The occupation number for each level will be written as a 3-digit number. A number **216** would denote that there are 2 particles in ϵ_0, 1 in ϵ_1, and 6 in ϵ_2. We could list all possible 3-digit numbers between **000** and **999** in base 10. Many in this listing will include impossible cases, such as **217** which violates the condition that there are nine individual particles, so these extraneous cases must be eliminated. Nevertheless, we have a systematic way of considering all possibilities.

Here is another example. Assume there is only one particle with three different levels. Then, a 3-digit binary number can be used to specify all possibilities. The following three configurations would work: **100**, **010**, and **001**, but not others, such as **110**, **101**, or **111** which have more than 1 particle in total.

In addition to the constraint of a fixed total number of particles, there is one more constraint. In the previous example of splitting five $1 bills among three people, two quantities must remain constant throughout the distribution of the money: the total number of people and the total amount of money. This analogy is helpful when dealing with a multi-particle system with a fixed number of particles (i.e., number of people) and a fixed amount of total energy (i.e., total wealth). Let's imagine gas particles moving in a container at a constant temperature. In addition to the number of the gas particles, the total amount of energy shared by the gas particles is fixed if the container is isolated (no heat exchange with the environment) and rigid (no mechanical work done on the gas). This total energy may be shared in many ways, as individual gas particles swap energy through elastic collisions. The energy conservation or the constraint of the fixed total energy is also true for other isolated physical systems, such as electrons in a solid, where their occupation of quantum-mechanically determined discrete energy levels depends on the total available energy. Therefore, such constraints must be satisfied as we consider all possible configurations in the brute force algorithm.

```python
# Code Block 5.4

# Find the possible configurations in a brute-force manner,
# satisfying the number and energy constraints.

import math
import numpy as np

def num2base (num, base=10, max_num_digits=0):
    if num<=0: return np.array([0],dtype=int)
    num_digits = int(np.floor(math.log(num,base))+1)
    num_digits = np.max([num_digits,max_num_digits])
    n_new = num
    n = np.zeros(num_digits,dtype=int)
    for i in range(num_digits):
        j = num_digits - i - 1
        n[i] = np.floor(n_new/base**j)
        n_new = n_new - base**j*n[i] # remainder
    n = n[::-1] # reversed
    # check
    assert np.sum(n* (base**np.arange(num_digits))) == num
    return n

# Test the num2base() function with a few examples.
assert all(num2base(9,base=10)==np.array([9]))
assert all(num2base(9,base=3)==np.array([0,0,1]))
```

```
assert all(num2base(9,base=2)==np.array([1,0,0,1]))
assert all(num2base(9,base=10,max_num_digits=3)==np.array([9,0,0]))

# Check the simple case
# e.g., splitting 5 bills among 3 people or
# 3 particles occupying 6 energy levels
N = 3 # total number of people or particles
E = 5 # total amount of money or energy
n = list()
e = np.arange(0,E+1,1) # number of bills or energy levels
for k in range((N+1)**(E+1)):
    n_tmp = num2base(k,base=(N+1),max_num_digits=(E+1))
    if np.sum(n_tmp)==N: # check total number constraint.
        if np.sum(n_tmp*e)==E: # check total energy constraint.
            n.append(n_tmp)
n = np.array(n)
print(n)
# Note this method will take a really long time
# for large N or E.
```

```
[[0 1 2 0 0 0]
 [0 2 0 1 0 0]
 [1 0 1 1 0 0]
 [1 1 0 0 1 0]
 [2 0 0 0 0 1]]
```

The above brute force strategy of considering all possibilities and eliminating ones that do not satisfy the constraints is intuitive and straightforward. However, one serious drawback is its inefficiency, where the computational time increases exponentially. For example, if there are five individual particles with three different levels, we must consider all possible 3-digit numbers in base 6. Because each digit would take on a value between 0 and 5 in base 6, the total number of numbers to be considered is 6^3 (from 000 to 555), and all these numbers must be checked to see if they satisfy the constraining conditions. If there are ten particles and three different levels, we would consider 11^3 numbers, etc. Similarly, if we consider five particles with six different levels, there are 6^6 possibilities. If there are five particles with 12 different levels, there are 6^{12} possibilities. In other words, the number of possibilities to consider and the computational time increase rapidly.

5.7 CODE FOR ENUMERATING ALL POSSIBILITIES (VERSION 2)

The second computational routine performs the same job of listing all permutations but uses a powerful technique called recursion. The main idea is that we create a general function (called **perm_recursive()** below) that can be called within itself, but when it is called "recursively," it considers a smaller range of possible permutations.

For example, suppose there are nine particles with three levels ($\epsilon_0, \epsilon_1, \epsilon_2$). We can assign one particle to one of the three levels and then consider possible permutations with eight particles in the next recursive call of the same function **perm_recursive()**. Within this recursive call for eight particles, the function will again assign one particle to a particular energy level and make yet another recursive call with seven particles, and the process continues. In addition to considering fewer particles in the subsequent function calls, each recursive step considers a smaller amount of energy because some energy was taken up at the previous step. Once the function reaches a base case, where no particles are left to assign to an energy level, the recursive call stops.

Compare the time of running the following and the previous code blocks. You will notice that this recursive method is much faster while producing the same answer as the brute force method.

```
# Code Block 5.5

# Consider a more powerful method.
import numpy as np

def perm_recursive (e,N,E):
    # n = d-dim vector
    # e = d-dim vector
    # sum(n) = N, sum(e*n) = E
    # e(0) = 0
    # call perm with N-1 and with less E.

    assert E>=0
    assert N>=0
    assert all(e>=0)
    N = int(N)
    E = int(E)
```

```
        dim = len(e) # dimension of vectors (number of energy levels)
        if (N==0): # base case.
            if E==0: return np.zeros((1,dim)) # Solution found.
            else: return np.zeros((0,dim)) # No solution.

        n = np.zeros((0,dim))
        for i in range(dim):
            if (E-e[i])>=0: # enough energy to drill down recursively.
                n_next = perm_recursive(e,N-1,E-e[i])
                if len(n_next)>0: # Solution(s) was found.
                    n_next[:,i] = n_next[:,i]+1
                    n = np.vstack((n,n_next)) # Keep adding solutions
        return n

N = 3
E = 5
e = np.arange(0,E+1,1)

n = perm_recursive(e,N,E)

# remove duplicate solutions.
n, counts = np.unique(n,axis=0,return_counts=True)

# Check the number and energy constraints of all solutions.
assert all(np.sum(n,axis=1)==N)
assert all(np.dot(n,e)==E)

print(np.array(n).astype(int))
```

```
[[0 1 2 0 0 0]
 [0 2 0 1 0 0]
 [1 0 1 1 0 0]
 [1 1 0 0 1 0]
 [2 0 0 0 0 1]]
```

The variable **n** in the above code blocks is a two-dimensional array whose column corresponds to the number of individual particles in each level. Hence, the sum of each row is equal to N, satisfying the particle number conservation constraint. The average along each column gives the average occupancy of each level. It turns out that no matter what N and U are, lower energy levels are more likely to be occupied than the higher levels. This trend arises from the fact that there are many more ways to share U among N entities when most individuals take a small portion of U. We had seen this effect in an earlier chapter when we simulated the kinetic theory of gas. As gas particles divided up the energy randomly, the equilibrium distribution looked like an exponential decay. As shown in Figure 5.3, the plot of average occupancy shows the same trend. We can obtain the same exponential behavior by considering all possible ways of distributing a fixed amount of energy among a fixed number of particles, as shown below.

```python
# Code Block 5.6

# Draw average occupancy graph of the multi-particle system.

N = 10
E = 5
e = np.arange(0,E+1,1)

n = perm_recursive(e,N,E)
n, counts = np.unique(n,axis=0,return_counts=True)

# Check the number and energy constraints of all solutions.
assert all(np.sum(n,axis=1)==N)
assert all(np.dot(n,e)==E)

print('All possibilities:')
print(np.array(n).astype(int))

print('Bar graph of average occupancy')
plt.bar(np.arange(0,E+1),np.mean(n,axis=0),color='k')
plt.ylabel('Average Occupancy')
plt.xlabel('E')
plt.savefig('fig_ch5_avg_occupancy.eps')
plt.show()
```

```
All possibilities:
[[5 5 0 0 0 0]
 [6 3 1 0 0 0]
 [7 1 2 0 0 0]
 [7 2 0 1 0 0]
 [8 0 1 1 0 0]
 [8 1 0 0 1 0]
 [9 0 0 0 0 1]]
```

Bar graph of average occupancy

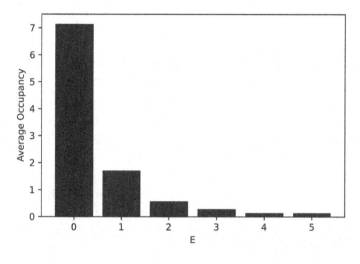

Figure 5.3

5.8 BOLTZMANN DISTRIBUTION

Now let's show mathematically how such an exponential distribution arises. We can start with $f(\epsilon)$, which represents the probability of finding a particle with specific energy ϵ. Let's consider two particular particles from a system of N particles with the total energy of U, and they would have the probabilities of $f(\epsilon_1)$ and $f(\epsilon_2)$ for having energies ϵ_1 and ϵ_2, respectively. Assuming these two particles are independent, the joint probability of this condition would be the product of individual probabilities, or $f(\epsilon_1)f(\epsilon_2)$.

Now, let's consider another system with $N - 1$ particles and the total energy of U. Since we usually deal with a system of many particles ($N \gg 1$), this new system with one less particle would be almost identical to the original system with N particles and U. Thus, the probability of a particular particle having energy $\epsilon_1 + \epsilon_2$ in the reduced system would be $f(\epsilon_1 + \epsilon_2)$. We may also treat the pair of particles in the original system with N particles as being a single particle in the new system with $(N - 1)$ particles. It is therefore expected that $f(\epsilon_1 + \epsilon_2)$ is equal to $f(\epsilon_1)f(\epsilon_2)$. We know that an exponential function, $e^{-\beta\epsilon}$, satisfies this relationship, because $e^{-\beta(\epsilon_1+\epsilon_2)} = e^{-\beta\epsilon_1}e^{-\beta\epsilon_2}$.

We can make a more rigorous justification of the exponential distribution using calculus, but let's briefly talk about a useful mathematical technique more generally.

5.9 MATH: LAGRANGE MULTIPLIER METHOD

We will work with a technique called the Lagrange Multiplier method. Here is an example: suppose we want to create a fence around an area using a limited amount of fencing material. If the total length of the fencing material is L, what is the maximum area A we can enclose? This problem can be phrased mathematically as the following: What is the optimal x and y, such that $A = xy$ is maximized, while satisfying a constraint, $2x + 2y = L$? We can say that we are maximizing $A(x, y)$, subject to $\phi(x, y) = x + y - L/2 = 0$. We can write a Lagrange function with a Lagrange multiplier, λ:

$$\mathcal{L}(x, y, \lambda) = A(x, y) + \lambda\phi(x, y) = xy + \lambda(x + y - L/2)$$

We set the partial derivatives with respect to x and y equal to zero and solve the resulting equations, including the constraint $\phi(x, y) = 0$.

$$\frac{\partial \mathcal{L}}{\partial x} = y + \lambda = 0$$

$$\frac{\partial \mathcal{L}}{\partial y} = x + \lambda = 0$$

$$x + y - L/2 = 0$$

With three equations and three unknowns (x, y, λ), the system of equations is solvable, yielding a solution of $x = y = L/4$ and the maximum area of $L^2/16$. That is, making a square enclosure gives the maximum area.

5.10 MATH: STIRLING'S APPROXIMATION

Another mathematical relationship we will be using is Stirling's approximation. It provides an approximate value of the factorial of a large number. Stirling's approximation states that for large n,

$$\ln n! \approx n \ln n - n.$$

It is straightforward to test the validity of the approximation, as shown in the following code block.

```python
# Code Block 5.7

# Test Stirling's approximation.

import numpy as np
import matplotlib.pyplot as plt

n_range = np.arange(5,20)
for n in n_range:
    v1 = np.log(np.math.factorial(n))
    v2 = n*np.log(n) - n
    print("For n =%3d: ln(n!) = %4.1f, nln(n)-n = %4.1f"%(n,v1,v2))
    perc_diff = (v1-v2)/v1*100
    plt.scatter(n,perc_diff,color='black')
plt.ylabel('Perc. Difference')
plt.xlabel('n')
plt.title("Goodness of Stirling's Approximation")
plt.savefig('fig_ch5_stirling_goodness.eps')
plt.show()
print("Percent Difference = (Actual-Approx)*100/Actual.")
print("The approximation gets better with large n.")
```

```
For n =  5: ln(n!) =  4.8, nln(n)-n =  3.0
For n =  6: ln(n!) =  6.6, nln(n)-n =  4.8
For n =  7: ln(n!) =  8.5, nln(n)-n =  6.6
For n =  8: ln(n!) = 10.6, nln(n)-n =  8.6
For n =  9: ln(n!) = 12.8, nln(n)-n = 10.8
For n = 10: ln(n!) = 15.1, nln(n)-n = 13.0
For n = 11: ln(n!) = 17.5, nln(n)-n = 15.4
For n = 12: ln(n!) = 20.0, nln(n)-n = 17.8
For n = 13: ln(n!) = 22.6, nln(n)-n = 20.3
For n = 14: ln(n!) = 25.2, nln(n)-n = 22.9
For n = 15: ln(n!) = 27.9, nln(n)-n = 25.6
For n = 16: ln(n!) = 30.7, nln(n)-n = 28.4
For n = 17: ln(n!) = 33.5, nln(n)-n = 31.2
For n = 18: ln(n!) = 36.4, nln(n)-n = 34.0
For n = 19: ln(n!) = 39.3, nln(n)-n = 36.9
```

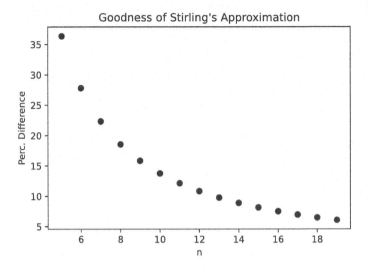

Figure 5.4

```
Percent Difference = (Actual-Approx)*100/Actual.
The approximation gets better with large n.
```

As a way of showing why this approximation works, we first note:

$$\ln n! = \ln n + \ln(n-1) + \ldots + \ln 2 + \ln 1 = \sum_{i=1,2,\cdots,n} \ln(i).$$

The above series is the sum of the areas of n rectangles whose widths are one and heights are $\ln(i)$, where i takes on values from 1 to n. Thus, the series can be considered an approximation of an integral of a natural log function from 1 to n, as illustrated by Figure 5.5.

```
# Code Block 5.8

n_max = 22
n = np.arange(1,n_max,1)
n_smooth = np.arange(1,n_max,0.1)
plt.plot(n_smooth,np.log(n_smooth),color='black')
plt.legend(('ln x',),framealpha=1.0)
plt.bar(n,np.log(n),width=0.8,color='gray')
plt.xticks((0,10,20))
plt.savefig('fig_ch5_stirling_approx.eps')
plt.show()
```

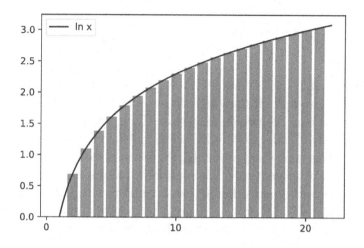

Figure 5.5

Therefore,

$$\ln n! \approx \int_1^n \ln x \, dx = (x \ln x - x)\Big|_1^n = n \ln n - n + 1.$$

Also, since n is much larger than 1, $-n + 1 \approx -n$, leading to the final expression of

$$\ln n! \approx n \ln n - n.$$

5.11   BACK TO THE BOLTZMANN DISTRIBUTION

Now to prove the Boltzmann distribution, we start with a function $\omega(n_0, n_1, \ldots)$, which denotes the number of configurations, or microstates, of having n_0 particles with energy ϵ_0, n_1 particles with energy ϵ_1, etc. Throughout this chapter, we have developed an intuition about ω. Suppose we have N distinguishable particles and need to choose n_0 particles to place into an energy level for ϵ_0; there are $\binom{N}{n_0} = \frac{N!}{(N-n_0)!n_0!}$ possible combinations. Then, we have n_1 particles out of the remaining $(N - n_0)$ for the ϵ_1 level, giving $\binom{N-n_0}{n_1} = \frac{(N-n_0)!}{(N-n_0-n_1)!n_1!}$ combinations. The product of these two numbers gives $\binom{N}{n_0}\binom{N-n_0}{n_1} = \frac{N!}{n_0!n_1!(N-n_0-n_1)!}$.

Continuing on with n_2, n_3, etc., we have an expression for $\omega(n_0, n_1, \ldots)$.

$$\omega(n_0, n_1, \ldots) = \frac{N!}{n_0! n_1! \ldots} = N! \prod_{i=0,1,\ldots} \frac{1}{n_i!}.$$

For more generality, let's assume that each energy level may have degeneracy g_i. For example, if there is only one way to have energy ϵ_4, then $g_4 = 1$, but if there are two distinct states for ϵ_4, $g_4 = 2$. As an analogy, we may picture a high-rise building with multiple floors, corresponding to energy levels, and there are multiple rooms or sections on each floor, corresponding to degeneracy. We could also picture different wealth or income levels, with many different career options with the same income. A particular orbital for an electron can accommodate an electron with spin up or down, so there is a two-fold degeneracy.

Imagine that six people ($n_7=6$) are sent to the seventh floor (ϵ_7), where 10 empty office spaces ($g_7=10$) are available. They are allowed to choose an office at random. The number of all possible ways of distributing six people into ten possible offices is $g_i^{n_i} = 10^6$. An unfortunate case will be that of all six people cramming into a single office space. Therefore, the above expression for $\omega(n_0, n_1, \ldots)$ needs to account for the extra possibilities arising from each level's degeneracy. The final expression is:

$$\omega(n_0, n_1, \ldots) = N! \prod_{i=0,1,\ldots} \frac{g_i^{n_i}}{n_i!}.$$

Note that if the particles are indistinguishable, the overall permutation factor $N!$ can be divided out. This extra constant factor does not change our derivations below.

Our goal is then to find a set of $n_0^*, n_1^*, \ldots$ that maximize ω, since they would be the microstate that is most probable and would be manifested at equilibrium. To find these optimal occupancy values, we use the Lagrange Multiplier method.

$$\mathcal{L}(n_0, n_1, \ldots; \alpha, \beta) = \ln \omega + \alpha \phi - \beta \psi,$$

where α and β are two Lagrange Multipliers, and $\phi = \sum_i n_i$ and $\psi = \sum_i n_i \epsilon_i$, as the total number of particles and total amount of energy are fixed. We also note that maximizing ω is equivalent to maximizing $\ln \omega$.

Using the additive property of logs and Stirling's approximation, we have

$$\ln \omega = \ln N! + \sum_i (n_i \ln g_i - n_i \ln n_i + n_i),$$

so

$$\frac{\partial \ln \omega}{\partial n_i} = \ln g_i - \ln n_i - 1 + 1 = \ln \frac{g_i}{n_i}.$$

Following the recipe of the Lagrange Multiplier technique, we obtain:

$$\frac{\partial \mathcal{L}}{\partial n_i} = \frac{\partial \ln \omega}{\partial n_i} + \alpha \frac{\partial \phi}{\partial n_i} - \beta \frac{\partial \psi}{\partial n_i} = \ln \frac{g_i}{n_i} + \alpha - \beta \epsilon_i = 0.$$

The above expression leads to

$$\ln \frac{n_i}{g_i} = \alpha - \beta \epsilon_i,$$

or

$$n_i = g_i e^{\alpha} e^{-\beta \epsilon_i},$$

which is the exponential behavior that we have been looking for.

## CHAPTER 6

Revisiting Ideal Gas

6.1 A LITTLE BIT OF QUANTUM MECHANICS

Quantum mechanics gives a different way to think about the ideal gas. Individual particles in an ideal gas can be described with a wavefunction representing the probabilistic nature of their existence. Let's start with a one-dimensional example, where the wavefunction $\psi(x)$ satisfies the time-independent Schrödinger's equation:

$$-\frac{\hbar^2}{2m}\frac{d^2\psi(x)}{dx^2} + V(x)\psi(x) = \epsilon\psi(x).$$

A particle with mass m contained within a specified volume with energy ϵ can be modeled as being subject to a potential $V(x)$ which is zero inside a specified range (between 0 and L) and infinite outside, so $\psi(x \leq 0) = 0$ and $\psi(x \geq L) = 0$. The corresponding wavefunction for the particle inside satisfies the following differential equation and the boundary conditions:

$$-\frac{\hbar^2}{2m}\frac{d^2\psi(x)}{dx^2} = \epsilon\psi(x),$$

$$\psi(0) = \psi(L) = 0.$$

The solutions can be written as the following:

$$\psi_n(x) = A\sin(k_n x),$$

where $k_n = \frac{n\pi}{L}$ for $n = 1, 2, 3, \ldots$.

The normalization condition determines the amplitude A. By plugging the solution back into Schrödinger's equation, we can find the energy of the particle to be

$$\epsilon(n) = \frac{\hbar^2 k_n^2}{2m} = \frac{h^2}{8mL^2}n^2,$$

which reveals the discrete nature of the energy levels, depending on the quantum number, n. Note $\hbar = \frac{h}{2\pi}$.

Let's visualize the first few energy levels. For convenience, we will assume $\frac{h^2}{8mL^2} = 1$ in the code.

```python
# Code Block 6.1

# Sketch the wavefunctions of a particle
# in an infinite well for the first 4 energy levels.

import numpy as np
import matplotlib.pyplot as plt

pi = 3.1415
L = 1
dx = 0.01
x = np.arange(0,L,dx)

for n in range(1,5):
    E = n**2
    kn = n*pi/L
    psi = np.sin(kn*x)
    psi = psi/np.sqrt(np.sum(psi**2)) # normalization
    # but the normalized wavefunction looks too short or tall,
    # so adjust the height of psi a bit (just for cosmetics).
    psi = psi*8

    plt.plot((0,L),(E,E),color='gray')
    plt.plot(x,psi+E,color='black',linewidth=3)
    plt.text(L+0.15,E,"n = %d"%n)

xbox_left = np.array([-0.1*L,0,0,-0.1*L])
ybox_left = np.array([0,0,E*1.1,E*1.1])

xbox_right = np.array([1.1*L,L,L,1.1*L])
ybox_right = np.array([0,0,E*1.1,E*1.1])

plt.fill(xbox_left,ybox_left,color='#CCCCCC')
plt.fill(xbox_right,ybox_right,color='#CCCCCC')
```

```
plt.plot((0,0),(0,E*1.1),color='gray')
plt.plot((L,L),(0,E*1.1),color='gray')
plt.ylim((0,E*1.1))
plt.xlabel('Position')
plt.ylabel('Energy')
plt.axis('off')
plt.savefig('fig_ch6_wavefunc.eps')
plt.show()
```

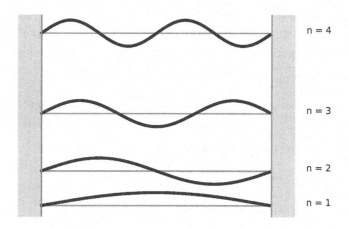

Figure 6.1

6.2 DEGENERACY

The above calculation has shown that a particle in a confined space (an infinite well) has discrete, quantized energy levels, and only those energies are allowed. If we expand this model into a three-dimensional space, we would consider a three-dimensional wavefunction $\psi(x, y, z)$ confined in a cube of side L.

The total energy now would be specified by three quantum numbers, (n_x, n_y, n_z) and $L = V^{1/3}$, so that

$$\epsilon(n) = \frac{h^2}{8mL^2}(n_x^2 + n_y^2 + n_z^2) = \frac{h^2}{8mV^{2/3}}n^2.$$

Then, we run into cases where the quantum state of a particle may be different even when the total energy is the same. For example, three different quantum states are possible for a single value of $\epsilon = h^2/(8mV^{2/3})$: with $n_x = 1, n_y = n_z = 0$, or with $n_y = 1, n_x = n_z = 0$, or with $n_z = 1, n_x = n_y = 0$. As an analogy, consider a building with many different rooms on different floors. A person may have a different amount of gravitational potential energy on different floors, but there are also different locations on the same floor with the same energy. As briefly introduced in the previous chapter, we call different states with the same energy "degenerate." The following code counts the number of degenerate cases for different amounts of energy, revealing that the degeneracy on average increases with the energy in the case of an ideal gas.

```
# Code Block 6.2

# Let's count the degenerate cases for different energy levels.

import numpy as np
import matplotlib.pyplot as plt

# E = total energy
# i, j, k = quantum number nx, ny, nz
# g = number of degenerate states

def break_into_sum_square (E,verbose=False):
    g = 0
    Emax = int(np.sqrt(E))+1
    for i in range(1,Emax):
        for j in range(1,Emax):
            for k in range(1,Emax):
                if i**2+j**2+k**2 == E:
                    g = g+1
                    if verbose:
                        print("%d^2 + %d^2 + %d^2 = %d"%(i,j,k,E))
    if verbose: print("Degeneracy = %d\n"%g)
    return g

break_into_sum_square(3,verbose=True)
break_into_sum_square(99,verbose=True)
break_into_sum_square(101,verbose=True)

assert break_into_sum_square(3)==1
assert break_into_sum_square(9)==3
```

```
1^2 + 1^2 + 1^2 = 3
Degeneracy = 1
```

```
1^2 + 7^2 + 7^2 = 99
3^2 + 3^2 + 9^2 = 99
3^2 + 9^2 + 3^2 = 99
5^2 + 5^2 + 7^2 = 99
5^2 + 7^2 + 5^2 = 99
7^2 + 1^2 + 7^2 = 99
7^2 + 5^2 + 5^2 = 99
7^2 + 7^2 + 1^2 = 99
9^2 + 3^2 + 3^2 = 99
Degeneracy = 9

1^2 + 6^2 + 8^2 = 101
1^2 + 8^2 + 6^2 = 101
2^2 + 4^2 + 9^2 = 101
2^2 + 9^2 + 4^2 = 101
4^2 + 2^2 + 9^2 = 101
4^2 + 6^2 + 7^2 = 101
4^2 + 7^2 + 6^2 = 101
4^2 + 9^2 + 2^2 = 101
6^2 + 1^2 + 8^2 = 101
6^2 + 4^2 + 7^2 = 101
6^2 + 7^2 + 4^2 = 101
6^2 + 8^2 + 1^2 = 101
7^2 + 4^2 + 6^2 = 101
7^2 + 6^2 + 4^2 = 101
8^2 + 1^2 + 6^2 = 101
8^2 + 6^2 + 1^2 = 101
9^2 + 2^2 + 4^2 = 101
9^2 + 4^2 + 2^2 = 101
Degeneracy = 18
```

```
# Code Block 6.3

# Now look at over a big range.

E_range = range(1000)
g_range = np.zeros(len(E_range))
for E in E_range:
    g_range[E] = break_into_sum_square(E)

plt.scatter(E_range,g_range,color='gray',s=3)
plt.xlabel('E')
plt.ylabel('Degeneracy')
plt.savefig('fig_ch6_degeneracy_scatter.eps')
plt.show()
```

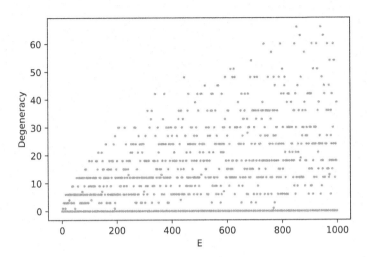

Figure 6.2

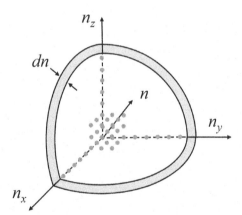

Figure 6.3

This counting process can be approximated by imagining an eighth of a sphere. Imagine a sphere in a positive octant that encloses integer lattice points. You might recall the spherical shell figure in Chapter 3, but now with three axes replaced by positive integer n_x, n_y, and n_z. The number of degenerate states with the same energy would equal the number of lattice points intersected by the sphere with the same radius n, where $n^2 = n_x^2 + n_y^2 + n_z^2$.

To make this argument more formal, let's define $N(n)$ as the number of integer lattice points enclosed within a sphere with radius n. Then $N(n)$ will be approximately equal to the volume of the sphere, $\frac{4}{3}\pi n^3$. We can define the density of lattice points as $g(n)$, so the number of lattice points between n and $n+dn$ would be $g(n)dn = N(n+dn) - N(n)$. Thus, $g(n) = \frac{dN(n)}{dn}$. Next, we make the identification that $n^2 = n_x^2 + n_y^2 + n_z^2 = \epsilon \frac{8mV^{2/3}}{h^2}$. With the change of variable $(n \to \epsilon)$,

$$n = \sqrt{\epsilon \frac{8mV^{2/3}}{h^2}}, \text{ so } \frac{dn}{d\epsilon} = \frac{1}{2\sqrt{\epsilon}} \sqrt{\frac{8mV^{2/3}}{h^2}}.$$

This leads to the mathematical expression for the degeneracy at ϵ as

$$g(\epsilon) = \frac{dN(n)}{dn}\frac{dn}{d\epsilon} = \frac{\pi}{4}\left(\frac{8mV^{2/3}}{h^2}\right)^{3/2}\epsilon^{1/2} = \frac{4\sqrt{2}\pi V m^{3/2}}{h^3}\epsilon^{1/2}.$$

We take an eighth of an entire sphere because the quantum numbers are positive integers only.

Figure 6.4 shows that this approximation is quite good.

```python
# Code Block 6.4

# Compare degeneracy and its moving average
# with the continuous approximation.

# continuous approximation
g_cont = (3.1415/4)*np.sqrt(E_range)
plt.scatter(E_range,g_range,color='gray',s=4)
plt.plot(E_range,g_cont,color='black',linewidth=3)
plt.xlabel('E')
plt.ylabel('Degeneracy')
legend_txt = ('Directly Counted','Continuous Approximation')
plt.legend(legend_txt,framealpha=1)
plt.savefig('fig_ch6_degeneracy_scatter_cont_approx.eps')
plt.show()

# moving average
window = 10
newE = len(E_range)-window
g_avg = np.zeros(newE)
E_avg = np.zeros(newE)

for i in range(newE):
    E_avg[i] = np.sum(E_range[i:i+window])/window
    g_avg[i] = np.sum(g_range[i:i+window])/window
# Note: subtract window/2, because we are averaging around a value.
E_avg = E_avg - window/2

plt.scatter(E_avg,g_avg,color='gray',s=3)
plt.plot(E_range,g_cont,color='black',linewidth=3)
plt.xlabel('E')
plt.ylabel('Degeneracy')
legend_txt = ('Moving Average','Continuous Approximation')
plt.legend(legend_txt,framealpha=1)
plt.savefig('fig_ch6_degeneracy_scatter_cont_approx_moving_avg.eps')
plt.show()
```

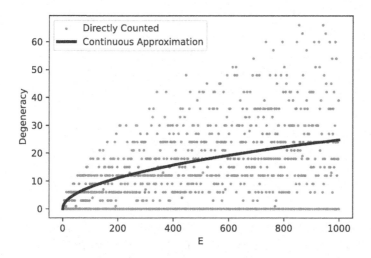

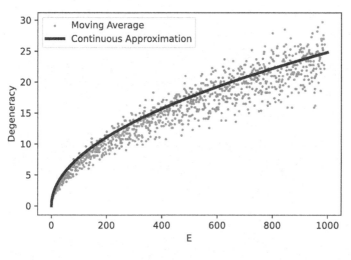

Figure 6.4

6.3 PARTITION FUNCTION

In mechanics and electrodynamics, a scalar function, potential, can reveal the direction and magnitude of force. Another scalar function, Lagrangian, can be used to derive the equation of motion of an object. In statistical mechanics, a scalar function known as a partition function, often denoted as Z, serves a similar role. It allows us to conveniently derive various thermodynamic quantities such as internal energy.

In the previous few chapters, we have worked with the key principle of statistical mechanics that a thermal system composed of many individual constituents produces behaviors that are calculatable and predictable using both energetic and entropic (i.e., probabilistic) considerations.

According to the Boltzmann distribution, the probability of being in the quantum state ϵ is proportional to:

$$P(\epsilon) = \frac{N(\epsilon)}{N} \propto g(\epsilon)e^{-\epsilon/k_B T}.$$

In earlier chapters, we used k for the Boltzmann distribution, but we will adopt a notation k_B to recognize it as the famous Boltzmann constant and to distinguish it from wavenumber $k_n = n\pi/L$.

The proportionality constant in the last expression is defined as $1/Z$, a reciprocal of the partition function. It can be determined by the normalization constraint $\int_0^\infty P(\epsilon)d\epsilon = 1$ (note $\epsilon \geq 0$).

In other words,

$$Z = \int_0^\infty g(\epsilon)e^{-\epsilon/k_B T}d\epsilon.$$

By putting the above expressions together with the expression for $g(\epsilon)$ for ideal gas, we have

$$Z = \int_0^\infty \frac{4\sqrt{2}\pi V m^{3/2}}{h^3}\epsilon^{1/2}e^{-\epsilon/k_B T}d\epsilon.$$

To simplify the integral, we can make a change of variable, $\frac{\epsilon}{k_B T} \to x$, so that the same integral can be written as the following:

$$Z = \alpha \int_0^\infty x^{1/2}e^{-x}dx,$$

where $\alpha = 4\sqrt{2}\pi \left(\frac{mk_B}{h^2}\right)^{3/2} VT^{3/2}$ and x is a unitless integration variable. We note that everything else other than V and T are just constants. The integral is finite and equal to $\frac{\sqrt{\pi}}{2}$, as will be shown in the following code blocks.

Therefore, we have the final expression for the partition function of ideal gas:

$$Z = \left(\frac{2\pi mk_B}{h^2}\right)^{3/2} VT^{3/2}.$$

Here we will show how to do a symbolic calculation using Python. We will calculate the integral of $x^{1/2}e^{-x}$ numerically and symbolically. The module **sympy** allows us to perform symbolic computation, which gives an accurate numerical value of 0.886226925452758, while the numerical calculation with **np.sum()** approaches this value for small enough integration step, **dx**.

```
# Code Block 6.5

# Numerical calculation.

import numpy as np
import matplotlib.pyplot as plt

dx = 0.001
x = np.arange(0,10,dx)
y = x**(0.5)*np.exp(-x)
plt.plot(x,y,color='k')
plt.xlabel('x')
plt.savefig('fig_ch6_integral_demo.eps')
plt.show()
# Area under the curve
print("Integral = %8.7f"%(np.sum(y)*dx))
```

Integral = 0.8860699

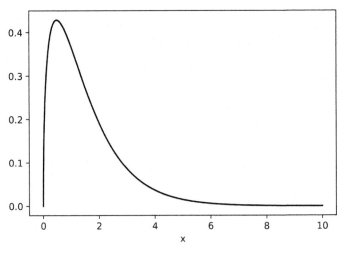

Figure 6.5

```
# Code Block 6.6

# Symbolic calculation ver. 1

import sympy as sym

x = sym.Symbol('x')
# sym.oo is symbolic constant for infinity.
sym.integrate(x**(0.5)*sym.exp(-x), (x,0,sym.oo))
```

0.886226925452758

Let's do one more clever change of variable, $x \to y^2$, so that

$$\int_0^\infty x^{1/2}e^{-x}dx = \int_0^\infty ye^{-y^2}(2ydy) = 2\int_0^\infty y^2e^{-y^2}dy.$$

When the latter integral is entered into **sympy** as shown below, we obtain the analytical solution $\frac{\sqrt{\pi}}{2}$.

```
# Code Block 6.7

# Symbolic calculation ver. 2

y = sym.Symbol('y')
sym.integrate(2*(y**2)*sym.exp(-y**2), (y,0,sym.oo))
```

$$\frac{\sqrt{\pi}}{2}$$

6.4 AVERAGE ENERGY OF AN IDEAL GAS

Let's consider another mathematical relation: a derivative of the natural log of the partition function with respect to temperature, $\frac{\partial \ln Z}{\partial T}$, where T is regarded as a variable of the partition function.

$$
\begin{aligned}
\frac{\partial \ln Z}{\partial T} &= \frac{\partial \ln Z}{\partial Z} \frac{\partial Z}{\partial T} \\
&= \frac{1}{Z} \frac{\partial Z}{\partial T} \\
&= \frac{1}{Z} \frac{\partial}{\partial T} \int_0^\infty g(\epsilon) e^{-\epsilon/k_B T} d\epsilon \\
&= \frac{1}{k_B T^2} \frac{1}{Z} \int_0^\infty \epsilon g(\epsilon) e^{-\epsilon/k_B T} d\epsilon.
\end{aligned}
$$

In the latter expression, we can identify that the integrand divided by Z is ϵ times the probability $P(\epsilon)$, so this can be further written as

$$
\frac{\partial \ln Z}{\partial T} = \frac{1}{k_B T^2} \int_0^\infty \epsilon P(\epsilon) d\epsilon.
$$

Because the average energy $<U> = \int_0^\infty \epsilon P(\epsilon) d\epsilon$, we conclude

$$
<U> = k_B T^2 \frac{\partial \ln Z}{\partial T}.
$$

In other words, the partition function allows us to calculate the average energy of a thermal system if its mathematical form is known. We can take its partial derivative with respect to T. Let's apply this result in the case of an ideal gas:

$$
\begin{aligned}
<U> &= k_B T^2 \frac{\partial \ln Z}{\partial T} \\
&= k_B T^2 \frac{\partial}{\partial T} \ln \left[\left(\frac{2\pi m k_B}{h^2} \right)^{3/2} V T^{3/2} \right] \\
&= k_B T^2 \left(\frac{3}{2} \right) \frac{\partial \ln T}{\partial T} \\
&= \frac{3}{2} k_B T.
\end{aligned}
$$

We just reproduced $< U >= \frac{3}{2}k_B T$, which is one of the main conclusions of the kinetic theory of gas. It is a significant and exhilarating result, as we have derived this relationship using two different approaches.

6.5   VISUALIZING ENERGY LEVELS WITH DEGENERACY

The following code block generates an example plot for visualizing energy levels with degeneracy. The vertical axis, as before, represents the energy levels, and the number of boxes along the horizontal axis corresponds to the number of degenerate states at each energy level. As we have discussed above, the higher energy levels tend to have more degenerate states, as illustrated in Figure 6.6.

```python
# Code Block 6.8

def sketch_occupancy_with_degeneracy (n):

    # Define the size of boxes
    marg = 0.1 # Size of margin
    h = 1.0-2*marg
    w = 1.0-2*marg
    xbox = np.array([marg,marg+w,marg+w,marg])
    ybox = np.array([marg,marg,marg+h,marg+h])

    N = len(n) # Number of energy levels
    max_g = 1 # Maximum number of degenerate states
    for each_level in n:
        max_g = np.max([max_g,len(each_level)])

    for i in range(N):
        for j in range(len(n[i])):
            plt.fill(xbox+j,ybox+i,color="#AAAAAA")
            x = (np.random.uniform(size=n[i][j])-0.5)*w*0.9+0.5+j
            y = (np.random.uniform(size=n[i][j])-0.5)*h*0.9+0.5+i
            plt.scatter(x,y,marker='.',color='k',s=50,zorder=2.5)

    plt.yticks([])
    plt.xticks([])
    plt.ylabel('Energy Levels')
    plt.axis('equal')
    plt.title("Occupancy:\n%s"%n)
    plt.box(on=False)

n = list([[5],[2,1,0],[0,0,1,2,0],[0,1,0,0,0,0,0]])
fig = plt.figure(figsize=(6,4))
```

```
sketch_occupancy_with_degeneracy(n)
plt.arrow(0, 0, 0, len(n)-0.1, head_width=0.05, head_length=0.1)
plt.savefig('fig_ch6_occupancy_with_degeneracy.eps')
plt.show()
```

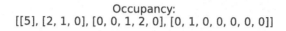

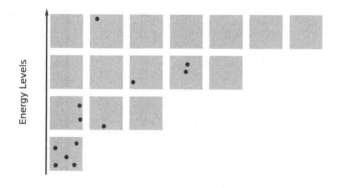

Figure 6.6

## CHAPTER 7

Revisiting Thermal Processes

7.1 REVIEW

The First Law of Thermodynamics is a statement of the energy conservation principle. Classically, it is often written as

$$dU = \delta Q - \delta W,$$

where U is the internal energy of a thermal system, such as ideal gas, and the change in U comes from either heat exchange Q or work done by the system W. (The symbol δ denotes the fact that Q and W are inexact differentials, which means that these variables are path-dependent. In other words, the details of the thermal processes, even when the beginning and end points might be the same, would affect the integrals of Q and W. For example, an ideal gas may go from a particular beginning condition, P_i and V_i, to an end condition, P_f and V_f, following two different processes: (1) an isothermal process or (2) an isochoric followed by an isobaric process.

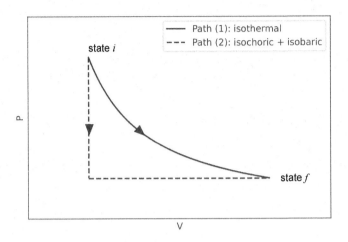

Figure 7.1

The total amounts of Q and W involved in these two different processes would be different, but the total change of internal energy, $U_f - U_i$, would be identical. In other words, we can change the internal energy of a thermal system in many different ways, and each way involves different combinations of Q and W.)

In statistical mechanics, there are N_i particles in the i-th energy level with energy ϵ_i, so that the total energy is given by the sum of the energy of all particles, $U = \sum_i N_i \epsilon_i$, which offers a different way to think about the change in internal energy:

$$dU = \sum_i dN_i \epsilon_i + \sum_i N_i d\epsilon_i.$$

In other words, any change in U may be due to the change in the occupancy of the energy levels or the change in the energy levels themselves. One might imagine a situation like this. There are N people in a building, and the total gravitational potential energy of these people may change when people move between floors (dN) or when an elevator moves vertically with people inside ($d\epsilon$).

These two expressions about dU give different insights about the energy of a thermal system. We may describe U as being affected by the heat

Q in/out-flux or work W done on/by the thermal system. Alternatively, we may view U as being changed from N_i and ϵ_i. When uniting these descriptions, heat influences N_i, while work influences the energy levels ϵ_i. For example, the total internal energy can be increased by either adding heat or performing work on the system. When heat is added, the particles in the lower energy levels are promoted to higher levels. When work is done adiabatically (without any heat exchange), the energy levels themselves rise, while the occupancy remains unchanged. More generally, a combination of these two mechanisms dictates the change of internal energy.

In this section, we will revisit the topic of thermal processes for ideal gas and examine them with the insights of statistical mechanics that we have developed in the previous chapters. We will use the results about the Boltzmann distribution where

$$\frac{N_i}{N} = \frac{g(\epsilon_i)e^{-\epsilon_i/k_BT}}{Z} \text{ (Occupancy Equation),}$$

where Z is a partition function and $g(\epsilon_i)$ is the degeneracy.

When we treat ideal gas as N indistinguishable quantum particles enclosed in a three-dimensional rigid box, we obtain the following results:

$$\epsilon_i = \frac{h^2}{8mV^{2/3}}(n_x^2 + n_y^2 + n_z^2) \text{ (Energy Equation),}$$

where n's are positive integers.

The degeneracy $g(\epsilon_i)$ of energy level can be approximated by the following expression:

$$g(\epsilon_i) = \frac{4\sqrt{2\pi}Vm^{3/2}}{h^3}\epsilon_i^{1/2} \text{ (Degeneracy Equation).}$$

The partition function for an ideal gas is

$$Z = \left(\frac{2\pi mk_B}{h^2}\right)^{3/2} VT^{3/2} \text{ (Partition Function).}$$

7.2 THERMAL PROCESSES

In the following numerical analysis, we will vary T and V. When we change T, the occupancy of different energy levels will change according to the above Occupancy Equation. As T increases, the lower energy levels are still favored over the higher ones, but the difference in the number of particles in different energy levels decreases, and the occupation numbers become more comparable in magnitude. When the gas is compressed and V decreases, the energy levels ϵ_i's increase according to the above Energy Equation. The degeneracy of each energy level would also change according to the above Degeneracy Equation. The following helper functions implement these equations. For numerical simplicity, we will assume constants k_B, m, and h have a value of 1 throughout all calculations.

```python
# Code Block 7.1

import numpy as np
import matplotlib.pyplot as plt

# Constants.
h = 1. # Planck constant
k_b = 1. # Boltzmann constant
m = 1. # mass of particle
pi = 3.141592

def e_n (V,n): # Energy Equation
    e = h**2/(8*m)*n**2/(V**(2/3))
    return e

def g_n (n): # Degeneracy for given n (in positive octant)
    g = (1/8)*4*pi*n**2 # density of lattice points
    return g

def g_e (V,e): # Degeneracy Equation
    g = np.zeros(len(e))
    g = 4*np.sqrt(2)*pi*m**(3/2)*V*e**(1/2)/h**3
    return g

def Z (V,T): # Partition Function
    Z = V*(2*pi*m*k_b*T/h**2)**(3/2)
    return Z
```

```
def P_n (V,T,n): # Occupancy Equation
    # Probability of occupancy
    e = e_n(V,n)
    bf = np.exp(-e/k_b/T) # Boltzmann distribution
    P = bf*g_n(n)/Z(V,T)
    return P
```

Note that n's (n_x, n_y, n_z) in the energy equation are positive integers that specify a quantum state of an individual particle, and $n^2 = n_x^2 + n_y^2 + n_z^2$. A good geometric picture is a three-dimensional sphere with radius n along with integer lattice points, as shown in Chapter 6. For a given T and V, each particle will occupy one of the states represented by the lattice points in the positive octant, or an eighth of a sphere. While implementing the helper functions above, we have also defined the degeneracy in terms of this quantum number n as $g(n) = 4\pi n^2/8$. Depending on your choice of variable (either energy ϵ or quantum number n), there are degeneracy functions: $g(\epsilon)$, g_e() or $g(n)$, g_n().

It is an interesting question whether more than one particle can occupy the same state (c.f., fermions versus bosons), but for our discussion, we will assume there are much more states than the number of particles even for large N, so it is highly unlikely that multiple particles will occupy the same state. This assumption is called the dilute gas limit.

Then, for a given T and V, we can calculate a profile of energy $\epsilon(n)$ with function e_n(). We can also examine a profile of $P(n)$, function P_n(), which determines the probability of occupancy according to the Boltzmann distribution. For each n, $\epsilon(n)P(n)$ represents energy density, and its integral $\int_0^\infty \epsilon(n)P(n)dn$ would be equal to the average internal energy. The function U_n(V,T,n) in the following code block calculates the total internal energy for given V, T, and n, by numerically summing the energy density: U = np.sum(P*e)*dn. n denotes the quantum states (e.g., n_x, n_y, n_z), which are different from the occupancy numbers N_i. Let's examine these for a particular value of T and V.

```
# Code Block 7.2

def U_n (V,T,n):
    # Numerically integrate energy density and get U.
    e = e_n(V,n)
    dn = n[1]-n[0]
    P = P_n(V,T,n)
    U = np.sum(P*e)*dn
    return U
```

```python
def plot_results (V_range,T_range,n,plot_filename=''):
    N = len(T_range)
    U_range = np.zeros(N)

    if N>1:
        color_range = np.linspace(0.8,0.2,N)
    else:
        color_range = np.zeros(N)

    fig, axes = plt.subplots(1,5,figsize=(8,3))
    for j in range(N):
        col = (color_range[j],color_range[j],color_range[j])
        T = T_range[j]
        V = V_range[j]
        e = e_n(V,n)
        U = U_n(V,T,n)

        axes[0].plot(T,V,'o',color=col)
        axes[1].plot(n,e,'-',color=col)
        axes[2].plot(n,P_n(V,T,n),'-',color=col)
        axes[3].plot(n,e*P_n(V,T,n),'-',color=col)
        axes[4].plot(T,U,'o',color=col)

    axes[0].set_xlabel('T')
    axes[0].set_ylabel('V')
    axes[0].set_xlim((0.0,2.5))
    axes[0].set_ylim((0.0,2.5))
    axes[1].set_xlabel('Quantum States (n)')
    axes[1].set_ylabel('$\epsilon(n)$')
    axes[2].set_xlabel('Quantum States (n)')
    axes[2].set_ylabel('$P(n)$')
    axes[3].set_xlabel('Quantum States (n)')
    axes[3].set_ylabel('$\epsilon(n) P(n)$')
    axes[4].plot(np.array([0,2]),3/2*np.array([0,2]),'k-')
    axes[4].set_xlabel('T')
    axes[4].set_ylabel('U')
    axes[4].set_xlim((0,2.1))

    plt.tight_layout()
    if len(plot_filename)>0:
        plt.savefig(plot_filename)
    plt.show()

n = np.arange(0,15,0.1)
T_range = [1]
V_range = [1]
plot_results(V_range,T_range,n,plot_filename='fig_ch7_singleTV.eps')
```

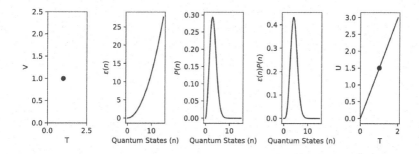

Figure 7.2

The above five plots require a close inspection. The point in the first plot specifies the state of an ideal gas on a V-vs.-T space. The next 3 plots show $\epsilon(n)$, $P(n)$, and $\epsilon(n)P(n)$ as a function of quantum number n. $\epsilon(n)$ shows the energy levels. $P(n)$ shows the Boltzmann distribution (i.e., higher energy levels are less likely to be populated), while considering the degeneracy and the size of the state space (i.e., lower quantum states are less likely to be populated because the number of states is small for low n), as we already observed from the Maxwell-Boltzmann distribution describing the speed of gas particles in Chapter 3. $\epsilon(n)P(n)$ shows the spectrum of energy density, whose integral is equal to the total internal energy, U. The point in the last plot specifies the state of ideal gas again, now on a U-vs.-T space. As expected, this point sits on a straight line of $U = \frac{3}{2}Nk_BT$.

Now let's examine the profiles of $\epsilon(n)$ and $P(n)$ for different thermal processes that cover different ranges of T and V, which are distinguished by contrasts of points and curves in each plot.

The first process is an isochoric process where V is held constant, as shown in Figure 7.3. We note that the energy levels, visualized by an $\epsilon(n)$-vs.-n graph, does not change. For higher T, the quantum states with higher n's are more likely to be occupied, as shown by the rightward-shifting curves in the middle. As a result, during an isochoric process of increasing T, the particles move from lower energy levels to higher levels, resulting in the overall increase of the internal energy. The work δW in the isochoric process is zero, so the heat δQ injected into the ideal gas is responsible for the change in U.

The second process, shown in Figure 7.4, is an adiabatic process where no heat enters or exits ($\delta Q = 0$), so that the occupancy of each quantum state does not change, while the energy levels themselves change. Therefore, the profiles of $P(n)$ are identical for various values of V and T along the adiabat described by PV^γ = constant or $TV^{\gamma-1}$ = constant. The increase in internal energy U at higher T comes from the elevation of the energy levels.

Imagine a population of people scattered within a high-rise building. The collective gravitational potential energy (like U) may be increased by people moving to the upper levels. A different situation would be everyone stays on the same floors, but each floor level rises mysteriously.

The third process we examine is an isothermal process, where T is held constant. In this case, both $\epsilon(n)$ and $P(n)$ change, but in such a manner that the integral of $\epsilon(n)P(n)$ stays constant. For example, as V increases, the energy level associated with each quantum state n decreases. However, the added heat energy promotes the gas particles in the lower quantum states with higher states. These two opposite trends are perfectly matched in the case of the isothermal process, so that the combined result is such that U remains constant, as shown in Figure 7.5.

```python
# Code Block 7.3
# Comparing different thermal processes.

# Case 1: de = 0 (or dW = 0)
print('V = const (isochoric process)')
print('e(n)-vs-n are the same.')
T_range = np.arange(0.5,2.1,0.25)
V_range = np.ones(len(T_range))
plot_results(V_range,T_range,n,plot_filename='fig_ch7_dVzero.eps')

# Case 2: dn = 0 (or dQ = 0)
# Change V according to PV**gamma = const = TV**(gamma-1)
print('Q = const (adiabatic process)')
print('P(n)-vs-n are the same.')
T_range = np.arange(0.5,2.1,0.25)
gamma = 5./3.
V_range = 1/T_range**(1/(gamma-1))
plot_results(V_range,T_range,n,plot_filename='fig_ch7_dQzero.eps')

# Case 3: dT = 0 (or dU = 0).
print('T = const (isothermal process)')
print('Integrals of e(n)P(n) are the same.')
V_range = np.arange(0.5,2.1,0.25)
T_range = np.ones(len(V_range))
plot_results(V_range,T_range,n,plot_filename='fig_ch7_dTzero.eps')
```

V = const (isochoric process)
e(n)-vs-n are the same.

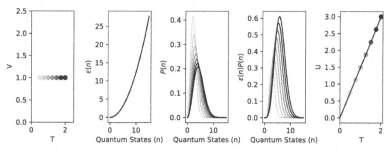

Figure 7.3

Q = const (adiabatic process)
P(n)-vs-n are the same.

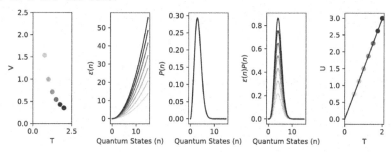

Figure 7.4

T = const (isothermal process)
Integrals of e(n)P(n) are the same.

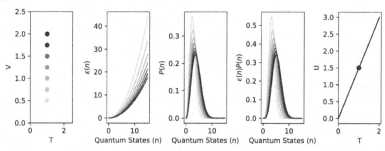

Figure 7.5

The above plotting exercise illustrates that a thermal process can be analyzed in terms of the occupancies and the energy levels of quantum states. Such analysis complements the classical viewpoints involving mechanical work that changes V and the heat flux during a thermal process.

7.3 CHECK

The following block of codes provides a couple of sanity checks of the earlier numerical routines. First, the function **P_n** creates $P(n)$, so that $\int_0^\infty P(n)dn = 1$, regardless of what T and V are. Second, the function **U_n** approximates $\int_0^\infty \epsilon(n)P(n)dn$, which should be equal to $\frac{3}{2}Nk_BT$. Given our numerical simplification of setting $k_B = 1$ and $N = 1$, **U_n** should return a value close to 1.5 for $T = 1$ and 3.0 for $T = 2$.

```
# Code Block 7.4


T = 1
V = 1
print('Normalization Check: Following values should be close to 1.0')
dn = 0.0001
n = np.arange(0,100,dn)
print("%.16f (for V=%f, T=%f) "%(dn*np.sum(P_n(V,2*T,n)),V, 2*T))
print("%.16f (for V=%f, T=%f) "%(dn*np.sum(P_n(V,2*T,n)),V, T))
print("%.16f (for V=%f, T=%f) "%(dn*np.sum(P_n(V,2*T,n)),2*V, T))
print("%.16f (for V=%f, T=%f) "%(dn*np.sum(P_n(V,2*T,n)),2*V, 2*T))
print('')
print('Total Energy Check: U = (3/2)*NkT')
print("U = %f (for V=%f, T=%f)"%(U_n (V,T,n),V,T))
print("U = %f (for V=%f, T=%f)"%(U_n (2*V,T,n),2*V, T))
print("U = %f (for V=%f, T=%f)"%(U_n (V,2*T,n),V, 2*T))
print("U = %f (for V=%f, T=%f)"%(U_n (2*V,2*T,n),2*V, 2*T))
```

```
Normalization Check: Following values should be close to 1.0
1.0000001040220627 (for V=1.000000, T=2.000000)
1.0000001040220627 (for V=1.000000, T=1.000000)
1.0000001040220627 (for V=2.000000, T=1.000000)
1.0000001040220627 (for V=2.000000, T=2.000000)

Total Energy Check: U = (3/2)*NkT
U = 1.500000 (for V=1.000000, T=1.000000)
U = 1.500000 (for V=2.000000, T=1.000000)
U = 3.000000 (for V=1.000000, T=2.000000)
U = 3.000000 (for V=2.000000, T=2.000000)
```

## CHAPTER 8

Entropy, Temperature, Energy, and Other Potentials

8.1 ENTROPY

In statistical mechanics, entropy S is defined as follows:

$$S = k_B \ln \omega.$$

It is proportional to the number of microstates ω, where the proportionality constant is the Boltzmann constant k_B. Entropy is one of the most fundamental concepts in thermal physics, and it helps us to understand how a thermodynamic state evolves over time. In a simple mechanical system like a ball placed on a landscape with hills and valleys, we can predict the ball's trajectory based on the analysis of its initial velocity and the shape of the landscape at its current position (i.e., the gradient). If there is a steep slope down the west side of a ball at rest, it will start rolling down in that direction, lowering its gravitational potential energy and gaining kinetic energy.

Similarly, we can predict how an isolated thermal system with fixed internal energy would evolve based on the analysis of its entropy because the system will move toward a state with higher entropy. In other words, the system tends toward a state with more possible configurations because a high-entropy state is more probable.

Let's revisit our old example of splitting up $5 among three individuals. We listed all possible permutations and showed that it is least likely for one person to have all the money because there is only one way to arrange such a situation. However, there are more ways to broadly distribute the five $1 bills among all people. We observed similar behavior when we simulated elastic collisions of gas molecules. Even if we started the simulation with one particle having all the energy (which is an unlikely situation), the total kinetic energy eventually gets shared among all particles (not uniformly but in an exponential form) because such a distribution is more likely. In an earlier chapter, we proved that the Boltzmann distribution maximizes the number of microstates and hence a thermal system will take on this state at equilibrium. Note that the system will continue to fluctuate dynamically about the final distribution, as the constituent particles will continue to exchange energy via existing interaction mechanisms.

Some people loosely describe entropy as a measure of "disorder," which is a reasonable but limited analogy. We might consider a system highly ordered when there are few ways of arranging its constituents. For example, a collection of coins is highly ordered and has low entropy if they are laid flat with all their heads facing upward. There is only one way to arrange the coins heads up. If half of the coins are facing up and the other half are facing down, as long as we do not care which particular coins are facing up, there are many more possible configurations, and the coin system is considered to have high entropy. The collection of the coins would look more disordered. A similar analogy can be applied to a collection of books in a library. The system has low entropy when the books are neatly ordered according to their assigned call numbers. There are many more ways of putting books on the shelves of a library if we disregard the call number system. Unless there is an active process or an agent to organize the library, the system will tend toward a high entropy state. Nevertheless, simply calling the entropy "disorder" does not fully capture the ideas of microstates and probability.

8.2 LAWS OF THERMODYNAMICS

There are four fundamental laws of thermodynamics.

The zeroth law states that heat flows from a hot body to a cold body when they are in thermal contact. Eventually, they will reach thermal equilibrium at the same temperature T.

The first law of thermodynamics states that the total energy is conserved so that it does not magically increase or decrease. Any change in the internal energy of a thermal system must be accounted for from its heat exchanges with the environment or from its mechanical work. This idea is often expressed as $dU = \delta Q + \delta W$, as we have seen in Chapter 4.

The second law of thermodynamics states that the entropy of an isolated thermal system never decreases. As we have discussed previously, it is a consequence of a higher-entropy state being more probable.

The third law of thermodynamics also deals with entropy and states that entropy approaches a constant value at the limit of zero temperature. For example, imagine an ideal gas approaching an absolute temperature of zero. Ignoring quantum mechanics momentarily, we would expect all gas particles are frozen still and occupy a single lowest energy level ($\omega = 1$). Its entropy will be a constant value of zero as $S = k_B \ln(1) = 0$, which is the smallest possible value for a thermal system. As an analogy, imagine a company experiencing a severe economic downturn, which must sell off all of its assets and all employees must share a single office in the basement of a building. The entropy of the company will be zero.

Now let's discuss how these three quantities, T, U, and S, are related.

8.3 TEMPERATURE AS A RATIO OF CHANGES IN ENERGY AND ENTROPY

Consider two systems in thermal contact, sharing a fixed total energy U. What would the number of microstates be for the joint system? Let $\omega_1(U_1)$ denote the number of microstates of the first system with energy U_1, and let $\omega_2(U_2)$ denote the same for the second system. Then, the number of microstates with U_1 and U_2 is $\omega_1(U_1)\,\omega_2(U_2)$. Because the energy of the first system U_1 can vary between 0 and U, while U_2 goes between U and 0, the total number of microstates is the sum of all possibilities:

$$\omega(U) = \sum_{U_1} \omega_1(U_1)\, \omega_2(U_2),$$

where $U_1 + U_2 = U$.

At thermal equilibrium, these two systems will have the same temperature, $T_1 = T_2 = T$, as a consequence of the zeroth law of thermodynamics. At thermal equilibrium, the most probable state, which is the state with the highest entropy or maximum $\omega(U)$, would have been reached, according to the second law of thermodynamics. The differential of $\omega(U)$ would be zero for infinitesimal energy exchange between the two systems. Mathematically,

$$d\omega = \left(\frac{\partial \omega_1}{\partial U_1}\right)\omega_2 dU_1 + \omega_1 \left(\frac{\partial \omega_2}{\partial U_2}\right) dU_2 = 0.$$

By dividing the above expression by $\omega_1 \omega_2$ and using $dU_1 + dU_2 = 0$, which is a consequence of the first law of thermodynamics, we obtain

$$\left(\frac{\partial \ln \omega_1}{\partial U_1}\right) = \left(\frac{\partial \ln \omega_2}{\partial U_2}\right).$$

According to Boltzmann's definition of entropy, $S = k_B \ln \omega$, the above relationship can be written as:

$$\left(\frac{\partial S_1}{\partial U_1}\right) = \left(\frac{\partial S_2}{\partial U_2}\right).$$

In other words, at thermal equilibrium, these two systems have the same temperature and the same value of $\frac{\partial S}{\partial U}$. Hence, T is intimately related to the ratio of changes in entropy and internal energy, while other state variables like V are fixed. Let's make the following definition for T:

$$T = \left(\frac{\partial U}{\partial S}\right)_V.$$

If we further allow the change in another state variable V, the change in internal energy dU can be split into two terms, where the first term captures the change due to dS and the second term is due to dV:

$$dU = \left(\frac{\partial U}{\partial S}\right)_V dS + \left(\frac{\partial U}{\partial V}\right)_S dV = TdS - PdV,$$

where the latter term is recognized as a contribution from mechanical work. This is another useful expression of the first law of thermodynamics. This formulation leads to yet another identification that

$$\delta Q_{\text{reversible}} = TdS.$$

The qualification of a "reversible" process in the above expression is a rather subtle point. It is related to the reason why we are using a symbol δ instead of the standard differential symbol d for Q and W. While the change in U only depends on the beginning and ending points of a thermal process, Q and W depend on the exact trajectory of a thermal process. This topic is beyond the scope of this book, and we would like to encourage readers to look further into it in other thermal physics books.

8.4 IDENTIFYING $\beta = 1/k_B T$

Continuing with the definition of entropy and the Boltzmann distribution, we will determine the Lagrange multiplier β as the inverse of $k_B T$. We have obtained the following results in Chapter 5:

$$n_i = g_i e^\alpha e^{-\beta \epsilon_i} \text{ (Boltzmann distribution)}$$

and

$$\omega(n_0, n_1, \ldots) = N! \prod_{i=0,1,\ldots} \frac{g_i^{n_i}}{n_i!} \text{ (Number of microstates).}$$

Now let's apply Boltzmann's definition of entropy and simplify the resulting expression with Stirling's approximation and some algebra.

$$
\begin{aligned}
S &= k_B \ln \omega \\
&= k_B \ln N! + k_B \sum_i \left(\ln g_i^{n_i} - \ln n_i! \right) \\
&= k_B \ln N! + k_B \sum_i \left(n_i \ln g_i - n_i \ln n_i + n_i \right) \\
&= k_B \ln N! + k_B \sum_i (-n_i) \left(\ln \frac{n_i}{g_i} - 1 \right) \\
&= k_B \ln N! + k_B \sum_i (-n_i) \left(\ln \left(e^\alpha e^{-\beta \epsilon_i} \right) - 1 \right) \\
&= k_B \ln N! + k_B \sum_i (-n_i)(\alpha - \beta \epsilon_i - 1) \\
&= k_B \ln N! - k_B(\alpha - 1) \sum_i n_i + k_B \beta \sum_i n_i \epsilon_i \\
&= k_B \left(\ln N! - (\alpha - 1)N \right) + k_B \beta U \\
&= S_o + k_B \beta U,
\end{aligned}
$$

where we lumped the three terms that do not depend on the energy as S_o.

When we take a partial derivative of entropy with respect to the energy, we arrive at a conclusion that $k_B \beta = \frac{\partial S}{\partial U}$. Furthermore, as we have defined T as a ratio of the change in internal energy and entropy, while other state variables, such as volume V and number of particles N, are held constant ($T = \frac{\partial U}{\partial S}$), we arrive at:

$$
\beta = \frac{1}{k_B T}.
$$

A later chapter on the two-state system discusses in detail how T influences the distribution of particles between two energy levels.

8.5 MATH: VOLUME OF A SPHERE

Let's take a quick digression and talk about a sphere in a d-dimension space with radius r. The volume of this hypersphere is given by:

$$V_d(r) = \frac{\pi^{d/2} r^d}{\Gamma\left(\frac{d}{2} + 1\right)}.$$

A special function, known as a Gamma function, has the following properties:

$$\Gamma\left(\frac{d}{2} + 1\right) = \begin{cases} k(k-1)\cdots(2)(1) & \text{if } d = 2k \\ (k - \frac{1}{2})(k - \frac{3}{2})\cdots(\frac{1}{2})\sqrt{\pi} & \text{if } d = 2k - 1 \end{cases}$$

where k is a positive integer.

The Gamma function has a more general definition, $\Gamma(z) = \int_0^\infty t^{z-1} e^{-t} dt$, where its value is defined for all complex numbers, except for non-positive integers. The factorial behavior of the Gamma function can be proven by performing integration by parts repeatedly. An extensive discussion of the Gamma function is beyond the scope of this book, so we will verify the above volume formula for a few values of d and refer the readers to other books on special functions. [†]

The following code block uses the **sympy** module to define the Gamma function, **gamma**. You can verify, for example, that the Gamma function evaluated at **d=6** is $\Gamma(6/2 + 1) = 3! = 6$, and at **d=8**, $\Gamma(8/2 + 1) = 4! = 24$, as expected. The code block applies this result to the spherical volume formula with $r = 1$ and obtains the volume of a unit circle in various dimensions. In particular, in three-dimensions (**d=3**), the volume formula correctly returns 4.189 since the volume of a sphere is $\frac{4}{3}\pi r^3 \approx 4.189 r^3$. A two-dimensional sphere is a circle, so for **d=2**, the volume formula correctly returns the value of 3.142 since the area of a circle is πr^2. A one-dimensional sphere is a line of length 2 that stretches between -1 and 1, so the volume formula for **d=1** returns 2, as expected.

[†] Brian Hayes wrote an interesting article about hypersphere titled, "An Adventure in the Nth Dimension," in American Scientist magazine. (www.americanscientist.org/article/an-adventure-in-the-nth-dimension)

```
# Code Block 8.1

# Obtain a symbolic expression for the Gamma function.
import sympy as sym
t = sym.Symbol('t')
z = sym.Symbol('z', positive=True)
gamma = sym.integrate(t**(z-1)*sym.exp(-t), (t,0,sym.oo))

dim_range = range(1,10,1)

print("Evaluate the Gamma function for a few sample values.")
for d in dim_range:
    # Substituting z with d/2+1.
    gamma_value = gamma.subs(z,d/2+1)
    print("d = %d, gamma(%2.1f) = %4.3f"%(d,d/2+1,gamma_value))

print("")
print("Evaluate the volume of a unit sphere in various dimensions.")
for d in dim_range:
    gamma_value = gamma.subs(z,d/2+1)
    vol_value = (3.1419**(d/2))/gamma_value
    print("d = %d, spherical volume = %4.3f"%(d,vol_value))
```

```
Evaluate the Gamma function for a few sample values.
d = 1, gamma(1.5) = 0.886
d = 2, gamma(2.0) = 1.000
d = 3, gamma(2.5) = 1.329
d = 4, gamma(3.0) = 2.000
d = 5, gamma(3.5) = 3.323
d = 6, gamma(4.0) = 6.000
d = 7, gamma(4.5) = 11.632
d = 8, gamma(5.0) = 24.000
d = 9, gamma(5.5) = 52.343

Evaluate the volume of a unit sphere in various dimensions.
d = 1, spherical volume = 2.000
d = 2, spherical volume = 3.142
d = 3, spherical volume = 4.189
d = 4, spherical volume = 4.936
d = 5, spherical volume = 5.265
d = 6, spherical volume = 5.169
d = 7, spherical volume = 4.726
d = 8, spherical volume = 4.060
d = 9, spherical volume = 3.300
```

We can further explore the topic of the volume of a hypersphere in a manner similar to calculating π in Chapter 1. In the following code block, we generate a large number (**N_total**) of random points between -1 and 1 along each dimension with **np.random.random()** function.

The d-dimensional coordinates are stored in variable `coord`, which is an array with `N_total` rows and `d` columns. We will count the number of points whose position is within 1 from the origin, `np.sum(dist<1)`, since these are the points within the unit sphere. The ratio of the number of the points inside of the unit sphere and the total number of points would approximate the ratio of volumes between a unit hypersphere and a hypercube with a side of 2 in a d-dimensional space.

$$\text{ratio} = \frac{N_{\text{inside sphere}}}{N_{\text{total}}} \approx \frac{V_{\text{sphere}}}{V_{\text{cube with side 2}}} = \frac{V_{\text{sphere}}}{2^d}.$$

Hence, we can obtain a reasonable approximation for a spherical volume by multiplying the number ratio by 2^d. The following code block and the resulting figure demonstrate this approach. There are a few interesting trends to note. As the dimensionality increases, the variation of the estimate increases or the precision decreases, since it would take exponentially more points to sample a higher dimension as densely. Hence, exploring a higher dimension becomes rapidly challenging and eventually impossible with a fixed resource. It is sometimes referred to as a "curse of dimensionality." Another interesting trend is that the volume of a unit sphere becomes vanishingly tiny as d increases, which is due to the fact that the corner regions of a hypercube, where a hypersphere cannot reach, grow very rapidly with the increasing dimensionality.

```
# Code Block 8.2

# Estimate the volume of a hypersphere.

import numpy as np
import matplotlib.pyplot as plt

N_trials = 10
N_total = 50000

volumes = np.zeros((N_trials,len(dim_range)))
for d in dim_range:
    for i in range(N_trials): # Multiple trials
        coord = np.random.random(size=(N_total,d))
        coord = coord*2 - 1 # Numbers are between -1 and 1.
        dist = np.sqrt(np.sum(coord**2,axis=1))
        ratio = np.sum(dist<1) / N_total
        volumes[i,d-1] = ratio*(2**d)

plt.boxplot(volumes)
plt.xlabel('Dimension')
```

```
plt.ylabel('Volume of Unit Sphere')
plt.ylim((0,7))
plt.xlim((0,10))
plt.xticks(dim_range)

dim_range_smooth = np.arange(0.5,9.5,0.1)
vol_value_smooth = np.zeros(len(dim_range_smooth))

for i,d in enumerate(dim_range_smooth):
    gamma_value = gamma.subs(z,d/2+1)
    vol_value = (np.pi**(d/2))/gamma_value
    vol_value_smooth[i] = vol_value
plt.plot(dim_range_smooth,vol_value_smooth,color='gray')
plt.savefig('fig_ch8_sphere_volume_dimension.eps')
plt.show()
```

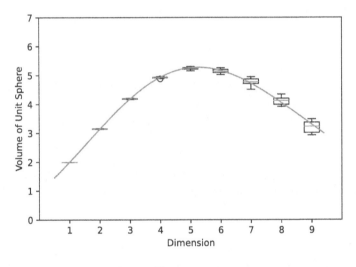

Figure 8.1

In this section, we did not derive the volume formula for a hypersphere and only demonstrated its plausibility. A rigorous proof and more discussions can be found in other books. In the following section, we will consider the volume of a hypersphere as a way of calculating the number of microstates. We will use $d = 3N$ and $r = n$. Since N is a large number, it does not particularly matter whether $3N$ is even or odd, so $\Gamma\left(\frac{3N}{2} + 1\right)$ will be written as $\frac{3N}{2}!$.

8.6 ENTROPY OF IDEAL GAS

In Chapter 6, we developed a strategy for counting the number of all possible states for a particle in a box. In this chapter, we will extend this strategy to derive an expression for the entropy of an ideal gas: N gas particles in a box of volume V. We need $3N$ quantum numbers since each particle would need one quantum number for each dimension. For example, particle 1 gets three quantum numbers, (n_{1x}, n_{1y}, n_{1z}), particle 2 gets (n_{2x}, n_{2y}, n_{2z}), and so on, until the last particle gets (n_{Nx}, n_{Ny}, n_{Nz}). Each quantum number is a positive integer. The set of all quantum numbers determines the energy U as $U = \frac{h^2}{8mV^{2/3}}n^2$ with $n^2 = \sum_{i=1}^{N} (n_{ix}^2 + n_{iy}^2 + n_{iz}^2)$.

Unlike the earlier case of a three-dimensional sphere for a single particle, we now have to deal with a $3N$-dimensional sphere, which is impossible to imagine or draw on paper. Nevertheless, the total number of available states can be approximated by a portion of a sphere's volume in a $3N$-dimensional space with a radius n. As you recall, when we worked with a single particle, we took an eighth of a three-dimensional sphere because the quantum numbers are positive. Likewise, for $3N$-dimensions, we multiply $\left(\frac{1}{2}\right)^{3N}$ to the volume of a sphere, $V_{3N}(n)$. With a formula for a spherical volume in a higher dimension discussed in the previous section, the total number of available states $\Omega(n)$ within n becomes:

$$\Omega(n) = \left(\frac{1}{2}\right)^{3N} V_{3N}(n) = \left(\frac{1}{2}\right)^{3N} \frac{\pi^{3N/2} n^{3N}}{\left(\frac{3N}{2}!\right)}.$$

Then, the number of microstates between n and $n + dn$, which corresponds to the number of lattice points embedded within a spherical shell of thickness dn, not within the full spherical volume, is:

$$\omega(n)dn = \Omega(n + dn) - \Omega(n).$$

Switching to a variable U instead of n, we have:

$$\omega(U) = \frac{d\Omega(n)}{dn}\frac{dn}{dU}$$

$$= \left(\frac{1}{N!}\right)\left(\frac{1}{2}\right)^{3N+1} \frac{3N\pi^{3N/2}}{\left(\frac{3N}{2}!\right)} \left(\frac{8mV^{2/3}}{h^2}\right)^{3N/2} U^{3N/2-1}.$$

Note we are dealing with indistinguishable particles, so an additional factor of $1/N!$ was introduced to avoid overcounting.

We can now take a natural log of $\omega(U)$ to find the entropy of an ideal gas. Upon going through a rather lengthy algebra and approximations based on large N, the entropy S of the ideal gas is:

$$S = k_B \ln \omega = k_B N \left\{ \frac{3}{2} \ln \left(\frac{4\pi m V^{2/3} U}{3 N^{5/3} h^2} \right) + \frac{5}{2} \right\}.$$

This result is known as the Sackur-Tetrode equation. If we calculate the rate of entropy change with respect to energy, $(\frac{\partial S}{\partial U})_V$, we obtain $\frac{3}{2}\frac{k_B N}{U}$, which also equals to $\frac{1}{T}$. It follows that $U = \frac{3}{2} N k_B T$, which is a major result for an ideal gas that we have encountered a few times already.

8.7 ENTROPY OF IDEAL GAS, AGAIN

Let's take a different approach to find the entropy of an ideal gas again. One of the main ideas from Chapter 5 was:

$$\omega(n_0, n_1, \cdots) = \prod_{i=0,1,\ldots} \frac{g_i^{n_i}}{n_i!},$$

where the factor of $N!$ was omitted with an assumption of indistinguishability. Now we apply the definition of entropy and use Stirling's approximation.

$$S = k_B \ln \omega$$
$$= k_B \sum_i (n_i \ln g_i - \ln n_i!)$$
$$= k_B \left(\sum_i n_i \ln g_i - \sum_i n_i \ln n_i + \sum_i n_i \right).$$

Let's work with the second term in the above expression by applying the Boltzmann distribution result, $\frac{n_i}{N} = \frac{g_i e^{-\epsilon_i/k_B T}}{Z}$.

$$\sum_i n_i \ln n_i = \sum_i n_i \ln \left(N \frac{g_i e^{-\epsilon_i/k_B T}}{Z} \right)$$

$$= \sum_i n_i \left(\ln N + \ln g_i + \ln e^{-\epsilon_i/k_B T} - \ln Z \right)$$

$$= \sum_i n_i \left(\ln N + \ln g_i - \left(\frac{\epsilon_i}{k_B T} \right) - \ln Z \right)$$

$$= N \ln N + \sum_i n_i \ln g_i - \frac{U}{k_B T} - N \ln Z,$$

where we used $\sum_i n_i = N$ and $\sum_i n_i \epsilon_i = U$ in the last step.

When we substitute the last expression back into the second term of the entropy, we obtain the following result:

$$S = k_B \left(\sum_i n_i \ln g_i - \left[N \ln N + \sum_i n_i \ln g_i - \frac{U}{k_B T} - N \ln Z \right] + N \right)$$

$$= k_B N \left(\ln \frac{Z}{N} + \frac{U}{N k_B T} + 1 \right).$$

This expression of S, thus far, is quite general and would apply to any thermal system at equilibrium. For an ideal gas specifically, we know $U = \frac{3}{2} N k_B T$ and have found an expression for the partition function of an ideal gas in Chapter 6:

$$Z = \left(\frac{4\pi m V^{2/3} U}{3 N h^2} \right)^{3/2}.$$

Thus, our final expression of entropy of an ideal gas is:

$$S = k_B N \left\{ \frac{3}{2} \ln \left(\frac{4\pi m V^{2/3} U}{3 N^{5/3} h^2} \right) + \frac{5}{2} \right\}.$$

It is identical to the entropy expression obtained by counting the microstates via the hypersphere's volume.

With this expression of S, let us plot the entropy of a monoatomic argon gas, an example of an ideal gas confined within a rigid box. In the following code block, we used the exact values of the fundamental constants to calculate the entropy of argon (Ar) gas while varying the temperature.

```
# Code Block 8.3

# Entropy of argon gas as a function of temperature.

import numpy as np
import matplotlib.pyplot as plt

m = 6.6*10**-26 # mass of Ar atom in [kg]
h = 6.63*10**-34 # Planck constant in [J sec]
pi = 3.14
N = 6.0*10**23 # Number of Ar particles (= Avogadro's number)
V = 10**-3 # Volume of container in [m^3]
dt = 0.0001 # Temperature step
T = np.arange(dt,300,dt) # Temperature in [K]
k = 1.38*10**-23 # Boltzmann constant in [J/K]
U =(3/2)*N*k*T # Energy in [J]

# Entropy in a unit of [J/K]
S = (3/2)*np.log((4*pi*m*(V**(2/3))*U)/(3*(N**(5/3))*(h**2)))+(5/2)
S = k*N*S
plt.plot(T,S,color='k')
plt.xlabel('T (K)')
plt.ylabel('S (J/K)')
plt.savefig('fig_ch8_entropy_Ar.eps')
plt.show()
```

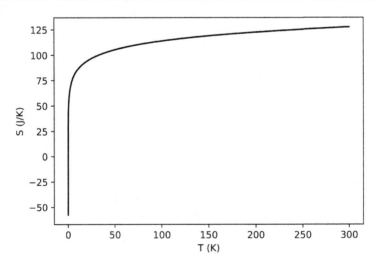

Figure 8.2

In the above S-vs.-T plot, we can see that the entropy diverges to $-\infty$ as T goes to zero, which seems consistent with the above expression for S, where $S \propto \ln U \propto \ln T$. However, this contradicts the third law of thermodynamics, which states that the entropy must approach a constant value at an absolute zero temperature.

This contradiction arises because our treatment of ideal gas has relied on a continuous approximation. Each discrete state of a gas particle was conceptualized as an integer lattice point within a phase space of quantum numbers. However, the total number of these quantum states was approximated by the continuous volume of a sphere in the phase space. As T approaches zero, the gas particles occupy the lowest energy state. The number of states available to the ideal gas decreases with decreasing T, but this number does not become zero. Imagine a sphere and the integral lattice points it encloses. As the volume of the sphere decreases, the number of enclosed lattice points decreases. However, even in the limit of zero volume, the number of enclosed lattice points is still one since an infinitesimal sphere would still include a point at the origin. Therefore, the continuous value of volume can not approximate the discrete number of lattice points well in this extreme limit.

8.8 MOTIVATION FOR OTHER METRICS OF A THERMODYNAMIC SYSTEM

Thus far in this book, we have worked extensively with internal energy U as an essential metric of a thermodynamic system. However, other related metrics become more useful under different conditions.

As a motivation, let's consider a simplistic example of measuring the value of a hypothetical company that sells a single product. The revenue of the company is calculated by $N_p V_p$, where N_p is the number of products sold and V_p is the price of the product. As a very crude model, the total value of the company may be given by $U_{\text{company}} = N_p V_p + U_o$, where U_o accounts for other quantities, such as its operating expenses and real estate values. The CEO of the company may be interested in the change in the company's value, $\Delta U_{\text{company}}$. If the company sells more products, the growth of the company can be calculated by $(\Delta N_p) V_p$. On the other hand, if the same number of products are sold at a higher unit price, the quantity of interest is $N_p (\Delta V_p)$.

We can make this example more interesting by adding the term $N_e V_e$, where N_e is the number of employees and V_e is the wage of each employee. Then, $H_{\text{company}} = U_{\text{company}} - N_e V_e$, where H_{company} is another function for describing the value of a company and it would be more useful than U_{company} if there have been any changes in the number of employees or their pay rate. Even with this simple example, there are potentially interesting interactions among the variables. For example, N_p and V_p may be inversely related since a customer may buy fewer products if they are too expensive, and V_p and V_e may be positively correlated since a high-value product may require high-wage employees to manufacture.

What about other factors? Can we develop a single "master" function that includes all possible factors and accounts for the raw value of a company in any situation? Maybe yes, theoretically, but no, practically. What would be interesting to a CEO is a small number of dynamic factors rather than other static factors that do not change over time. Therefore, the CEO and her team would work with different variables and various metrics that may be useful under different conditions.

Likewise, there are different economic indices, such as Gross Domestic Product (GDP) or Gross National Product (GNP). One economist might use GDP, the value of the finished domestic goods and services produced within a nation's borders, to measure a country's economic value. Another economist might use GNP, the value of all finished goods and services owned by a country's citizens, whether or not those goods are produced in that country. In finance, the indices like the S&P500, the Dow Jones Industrial Average, and the NASDAQ Composite in the US offer different ways of tracking and measuring the performance of a stock market or an economic sector. The choice of which stocks are included (and which ones are excluded) is an essential consideration for these indices. In thermal physics, too, we work with various metrics, or thermodynamic potentials with appropriate variables, to analyze different thermal systems or processes, as we will see below.

In our previous example with U_{company}, we discussed a pair of variables, (N_p, V_p), which are intimately related. Another pair of variables, (N_e, V_e), are closely related and always go together, too. For describing a thermal system, we will also deal with pairs of state variables, and each pair is known as a conjugate variable. Pressure P and volume V

make up a conjugate pair. Entropy S and temperature T form another conjugate pair. Another thermodynamic state variable is N, which denotes the total number of particles. So far it has been fixed, as we have only considered a closed thermal system where the number of particles is always constant. However, in an open thermal system where the particles may enter or exit the system, N becomes a variable quantity. A conjugate variable paired with N is called a chemical potential μ, and it denotes the change in energy accompanied by an addition of one particle into the thermal system.

As the temperature difference determines the direction of heat flow from high to low temperature, the chemical potential determines the direction of the particles' flow from a region of high chemical potential to a region of low chemical potential. The drive of chemical reactions or phase transitions is most naturally described in terms of chemical potential since the number of atoms or molecules belonging to a particular chemical species or phase changes during such processes.

Another interesting example where the concept of chemical potential is necessary is the case of Fermi energy. In a solid, electrons occupy energy levels similar to the particles in a box discussed in Chapter 6. According to the Pauli exclusion principle, each energy state can only accommodate a pair of electrons with opposite spins: up and down. These electron pairs will fill up the available energy states, starting from the lowest level. This situation can be imagined as a high-rise building where each floor can only accommodate two people and where all lower floors must be occupied before allowing people to move upstairs. Since there is a finite number of electrons in a solid, there will eventually be a top-level, up to which all levels are filled, and this is called the Fermi level. The associated maximum electron energy is called the Fermi energy, and it determines the electron distribution in a solid and characterizes its electronic structure. If we would like to add one extra electron to a solid, the only way is to place it at an energy level just above the Fermi level since there will not be any available spot below the Fermi level. Hence, the chemical potential of the new electron must be just above the Fermi energy. As we add more electrons, the chemical potential in a solid will correspondingly increase.

8.9 FOUR THERMODYNAMIC POTENTIALS: U, H, F, G

There are four thermodynamic potentials: internal energy U, enthalpy H, Helmholtz free energy F, and Gibbs free energy G. These are different metrics for describing a thermal system at any given moment. Let us start with the now familiar internal energy.

According to the first law of thermodynamics, the definition of internal energy U is given in a differential form as

$$dU = TdS - PdV + \mu dN.$$

This is the expression we saw before, plus a new term with the chemical potential and the particle number, hinting that the relevant variables involved in the change of internal energy are entropy, volume, and particle number. These are called natural variables and make the internal energy a function of S, V, and N. The total differential of $U(S, V, N)$ is

$$dU = \left(\frac{\partial U}{\partial S}\right)_{V,N} dS + \left(\frac{\partial U}{\partial V}\right)_{S,N} dV + \left(\frac{\partial U}{\partial N}\right)_{S,V} dN,$$

where the subscript symbols next to the right parenthesis indicate the state variables that are held constant for the partial derivatives.

As we compare these two expressions, we can make the following identifications, which are not new.

$$T = \left(\frac{\partial U}{\partial S}\right)_{V,N} \quad \text{(definition of temperature)}$$

$$P = -\left(\frac{\partial U}{\partial V}\right)_{S,N} \quad \text{(expression related to mechanical work)}$$

$$\mu = \left(\frac{\partial U}{\partial N}\right)_{S,V} \quad \text{(definition of chemical potential)}$$

Enthalpy H is defined as $U + PV$ and its differential is:

$$\begin{aligned} dH &= dU + (PdV + VdP) \\ &= (TdS - PdV + \mu dN) + (PdV + VdP) \\ &= TdS + VdP + \mu dN \end{aligned}$$

where dU is replaced with the differential expression from above. This tells us that for enthalpy, S, P, and N are the natural variables. This mathematical trick of swapping the role of a natural variable between a conjugate pair is called the Legendre transform. Now let's take a total differential of $H(S, P, N)$ with respect to these natural variables and discover how several thermal variables are related to partial derivatives of H:

$$dH = \left(\frac{\partial H}{\partial S}\right)_{P,N} dS + \left(\frac{\partial H}{\partial P}\right)_{S,N} dP + \left(\frac{\partial H}{\partial N}\right)_{S,P} dN.$$

$$T = \left(\frac{\partial H}{\partial S}\right)_{P,N}$$

$$V = \left(\frac{\partial H}{\partial P}\right)_{S,N}$$

$$\mu = \left(\frac{\partial H}{\partial N}\right)_{S,P}$$

Similarly, let's continue with Helmholtz free energy F, which is defined as $F = U - TS$.

$$dF = dU - (TdS + SdT)$$
$$= (TdS - PdV + \mu dN) - (TdS + SdT)$$
$$= -SdT - PdV + \mu dN.$$

For F, the natural variables are T, V, and N. Hence,

$$dF = \left(\frac{\partial F}{\partial T}\right)_{V,N} dT + \left(\frac{\partial F}{\partial V}\right)_{T,N} dV + \left(\frac{\partial F}{\partial N}\right)_{T,V} dN.$$

$$S = -\left(\frac{\partial F}{\partial T}\right)_{V,N}$$

$$P = -\left(\frac{\partial F}{\partial V}\right)_{T,N}$$

$$\mu = \left(\frac{\partial F}{\partial N}\right)_{T,V}$$

The last thermodynamic potential is Gibbs free energy G with a definition of $G = U - TS + PV$.

$$dG = dU - (TdS + SdT) + (PdV + VdP)$$
$$= (TdS - PdV + \mu dN) - (TdS + SdT) + (PdV + VdP)$$
$$= -SdT + VdP + \mu dN.$$

For G, the natural variables are T, P, and N.

$$dG = \left(\frac{\partial G}{\partial T}\right)_{P,N} dT + \left(\frac{\partial G}{\partial P}\right)_{T,N} dP + \left(\frac{\partial G}{\partial N}\right)_{T,P} dN.$$

$$S = -\left(\frac{\partial G}{\partial T}\right)_{P,N}$$

$$V = \left(\frac{\partial G}{\partial P}\right)_{T,N}$$

$$\mu = \left(\frac{\partial G}{\partial N}\right)_{T,P}$$

It is an instructive exercise to evaluate some of the above quantities for ideal gas. Starting with U and S for an ideal gas, we may define Helmholtz free energy and Gibbs free energy. Then, we can obtain an expression for pressure and chemical potential by taking an appropriate partial derivative.

We can verify that $P = -\left(\frac{\partial F}{\partial V}\right)_{T,N} = \frac{Nk_BT}{V}$, which is just an ideal gas law. We also obtain:

$$\mu = \left(\frac{\partial G}{\partial N}\right)_{T,P} = k_BT \ln\left(\frac{Nh^3}{V(2\pi mk_BT)^{3/2}}\right).$$

In the following code block, we demonstrate how Python's **sympy** module can recreate these results. **log** refers to a natural logarithm with base of e. It requires some straightforward algebra to verify that the above expression of μ is indeed equivalent to the symbolically-computed expression of **mu**.

```
# Code Block 8.4

import sympy as sym

# Calculate pressure from symbolic differentiation.

# Define symbols
k, T, N, V, h, m, pi = sym.symbols('k_{B} T N V h m \pi')

# Internal energy of ideal gas
U = (3/2)*N*k*T
# Entropy of ideal gas
S = (3/2)*sym.ln((4*pi*m*(V**(2/3))*U)/(3*(N**(5/3))*(h**2)))+(5/2)
S = k*N*S
# Helmhortz Free Energy
F = U-T*S

print('S for ideal gas')
display(S.nsimplify())
print('')

print('P from -(dF/dV) for ideal gas')
P = -sym.diff(F,V)
display(sym.nsimplify(P))
print('')

# Redefine P as a sympy symbol (override the previous definition)
P = sym.symbols('P')

# Gibbs Free Energy
G = U - T*S + P*V

# Calculate chemical potential from symbolic differentiation.
print('mu from (dG/dN) for ideal gas')
mu = sym.diff(G,N)
mu = mu.collect(k).collect(T)
display(sym.nsimplify(mu))
```

S for ideal gas

$$Nk_B\left(\frac{3\log\left(\frac{2TV^{\frac{2}{3}}\pi k_B m}{N^{\frac{2}{3}}h^2}\right)}{2}+\frac{5}{2}\right)$$

P from -(dF/dV) for ideal gas

$$\frac{NTk_B}{V}$$

mu from (dG/dN) for ideal gas

$$-\frac{3Tk_B \log\left(\frac{2TV^{\frac{2}{3}}\pi k_B m}{N^{\frac{2}{3}} h^2}\right)}{2}$$

8.10 THERMODYNAMIC RELATIONS

Let's take one more step and work with the second-order partial derivatives of the above expressions. We can obtain a few other useful equalities known as Maxwell's relations, whose derivations are based on the fact that a mixed second-order partial derivative, successive differentiation of a function with respect to two independent variables, remains identical regardless of the order of the differentiation. For example, consider a second-order partial differentiation of internal energy, first with respect to entropy and then with respect to volume:

$$\left(\frac{\partial}{\partial V}\left(\frac{\partial U}{\partial S}\right)_{V,N}\right)_{S,N} = \left(\frac{\partial T}{\partial V}\right)_{S,N}$$

where we used $\left(\frac{\partial U}{\partial S}\right)_{V,N} = T$.

Now let's make the order of differentiation reversed so that

$$\left(\frac{\partial}{\partial S}\left(\frac{\partial U}{\partial V}\right)_{S,N}\right)_{V,N} = -\left(\frac{\partial P}{\partial S}\right)_{V,N}$$

where we used $\left(\frac{\partial U}{\partial V}\right)_{S,N} = -P$.

Since these two second-order derivatives should be identical, we have

$$\left(\frac{\partial T}{\partial V}\right)_{S,N} = -\left(\frac{\partial P}{\partial S}\right)_{V,N}.$$

Similarly, we can calculate the mixed second-order derivatives for each of H, F, and G and have the followings:

$$\left(\frac{\partial T}{\partial P}\right)_{S,N} = \left(\frac{\partial V}{\partial S}\right)_{P,N}$$

$$\left(\frac{\partial S}{\partial V}\right)_{T,N} = \left(\frac{\partial P}{\partial T}\right)_{V,N}$$

$$-\left(\frac{\partial S}{\partial P}\right)_{T,N} = \left(\frac{\partial V}{\partial T}\right)_{P,N}$$

Let's check Maxwell's relations by calculating $\left(\frac{\partial S}{\partial V}\right)_{T,N}$ and $\left(\frac{\partial P}{\partial T}\right)_{V,N}$ for ideal gas separately.

```
# Code Block 8.5

import sympy as sym
k, T, N, V, h, m, pi = sym.symbols('k_{B} T N V h m \pi')

# U, S, and P for ideal gas.
U = (3/2)*N*k*T
S = (3/2)*sym.ln((4*pi*m*(V**(2/3))*U)/(3*(N**(5/3))*(h**2)))+(5/2)
S = k*N*S
P = N*k*T/V

print('dS/dV for fixed T and N')
dS_over_dV = sym.diff(S,V)
display(sym.nsimplify(dS_over_dV))

print('dP/dT for fixed V and N')
dP_over_dT = sym.diff(P,T)
display(sym.nsimplify(dP_over_dT))

print('Note that these two expressions are equal.')
```

dS/dV for fixed T and N

$$\frac{Nk_B}{V}$$

dP/dT for fixed V and N

$$\frac{Nk_B}{V}$$

Note that these two expressions are equal.

We have shown that $\left(\frac{\partial S}{\partial V}\right)_{T,N}$ is identical to $\left(\frac{\partial P}{\partial T}\right)_{V,N}$ as an illustration of one of the Maxwell's relations. These relations imply that if one knows the ratio of changes in one pair of variables, one can also find the ratio of changes in the other pair. For example, the rate of change in entropy with respect to volume at constant temperature and particle number can be determined by the rate of change in pressure with respect to temperature for fixed volume and gas particle number. The latter quantity can be experimentally measurable using barometers and thermometers, while the former quantity involving entropy may not be directly measurable.

It is interesting to note that many of the thermodynamic relations and definitions are expressed in terms of a rate of change between two variables or a partial derivative. We often define a function as a mapping between an input value x and the corresponding point-wise output $f(x)$. Our study of calculus shows an alternative way of dealing with a function. If we know the derivative $\frac{df(x)}{dx}$ at all points and a single value $f(x_0)$ at some reference point x_0, we can determine the point-wise value of a function by integration: $f(x) = \int_{x_0}^{x} \left(\frac{df(s)}{ds}\right) ds + f(x_0)$. That is, if we know $f(x)$, we can of course calculate $\frac{df}{dx}$, but what is interesting is that if we know $\frac{df}{dx}$ and $f(x_0)$, we can also determine $f(x)$.

The derivative or the rate of change can be more interesting and useful than the value of a function. For example, when we are hiking, what we tend to notice and find useful is how steep or shallow the local landscape is, rather than the exact height with reference to the sea level. In kinematics, the instantaneous velocity $v(t)$, a time rate of change in position, of a moving object gives more interesting information, such as its kinetic energy or momentum, than its exact position $x(t)$. Also knowing $v(t)$ allows us to calculate $x(t)$ by $x(t) = \int_{0}^{t} v(s)ds + x(0)$. The change may also be more readily and directly measurable than the raw values since the former does not require setting a reference value. For example, in classical mechanics, it is convenient to work with the change in gravitational potential energy of a falling object as $mg\Delta x$, but if you would like to work with the exact value of gravitational potential energy at a particular height, an additional definition about a reference point ($x = 0$) should be introduced. Thus, the utility of a thermodynamic potential, like gravitational or electrical potential, lies in its change during a thermodynamic process between different states, rather than its point-wise value at a particular state.

III

Examples

## CHAPTER 9

Two-State System

Imagine a town where there are only two-story apartments. Living on an upper floor unit is more expensive than living on a lower floor unit. We could survey the residents and find out where they live. It would not be surprising to find that the upper units are less occupied because they are more expensive. The higher the price difference between the upper and lower units is, the more significant the difference in the occupancies would be. However, the price difference would matter in the context of the average wealth of the occupants.

Mathematically, we denote the energy difference between the two states as $\Delta \epsilon = \epsilon_2 - \epsilon_1$, which would correspond to the price difference between the upper and lower units. The temperature multiplied by the Boltzmann constant, $k_B T$, would be analogous to the average wealth of the occupants. The occupancy of these two levels is determined by the Boltzmann distribution, and the ratio of the occupancy is

$$\frac{N_2}{N_1} = \frac{e^{-\epsilon_2/k_B T}}{e^{-\epsilon_1/k_B T}} = e^{-\frac{\Delta \epsilon}{k_B T}}.$$

When $\Delta \epsilon = 0$, there is no difference between the two states, so $N_1 = N_2$. For $\Delta \epsilon > 0$, $\frac{N_2}{N_1} < 1$, or N_2 is smaller than N_1. The difference between N_2 and N_1 is also affected by the value of $k_B T$. It can be interpreted as a representative energy scale or the average thermal energy of a thermal system at T. If the energy difference $\Delta \epsilon$ is significantly smaller than $k_B T$ (that is, the upper-level units are not that expensive), N_2 is slightly smaller than N_1. On the other hand, if $\Delta \epsilon$ is appreciably larger than $k_B T$, N_1 will be significantly larger than N_2. The following code block illustrates this trend by plotting $\frac{N_2}{N_1}$ versus $\frac{\Delta\epsilon}{k_B T}$ for two different values of T (low and high). Note $\frac{N_2}{N_1}$ will always be less than 1.

```
# Code Block 9.1
# Comparing the occupancy between two states.

import matplotlib.pyplot as plt
import numpy as np

kT_hi = 100
kT_lo = 1
e = np.arange(0,5,0.1)
plt.plot(e,np.exp(-e/kT_hi),color='k',linestyle='solid')
plt.plot(e,np.exp(-e/kT_lo),color='k',linestyle='dotted')
plt.legend(('High T','Low T'),framealpha=1)
plt.ylabel('$N_2 / N_1$')
plt.xlabel('$\Delta \epsilon / k_B T$')
plt.ylim((0,1.1))
plt.yticks(ticks=(0,0.5,1))
plt.savefig('fig_ch9_ratio_vs_epsilon.eps')
plt.show()
```

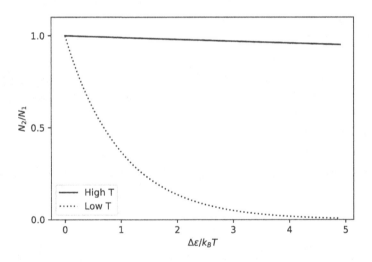

Figure 9.1

Let's create a function for visualizing the particle distribution, or occupancy, within a two-state system. This new function, sketch_distrib_2states(), builds upon the sketch_occupancy()

function from Chapter 5. In addition to the familiar boxes and dots representing the energy levels and the particles, there will be an additional bar graph overlaid with the exponential curve representing the Boltzmann distribution. Since we are mostly concerned with the energy difference between two states, we will assume that the lower state has zero energy and the higher state with $\Delta\epsilon$. By doing so, we can easily define the partition function for the two-state system as $Z = e^{-0/k_BT} + e^{-\Delta\epsilon/k_BT} = 1 + e^{-\Delta\epsilon/k_BT}$. With this, the probability or the fraction of particles at each level will be $\frac{N_1}{N_1+N_2} = \frac{e^{-0/k_BT}}{Z} = \frac{1}{Z}$ and $\frac{N_2}{N_1+N_2} = \frac{e^{-\Delta\epsilon/k_BT}}{Z}$, as we discussed in Chapter 6.

This dual representation allows us to visualize the current state of our thermal system in several ways. The particle distribution can be inferred intuitively from the density of dots of each box, with annotated values of the fraction of total particles on the left. The height of the bars in the bar graph on the right also shows the fraction of occupancy. The gray shading of the bars corresponds to the shading of the boxes, where the lower energy level is coded with a darker shade. Hence, the height of the darker bar is given by $\frac{N_1}{N_1+N_2}$, and the height of the lighter bar is $\frac{N_2}{N_1+N_2}$.

```
# Code Block 9.2

import matplotlib.pyplot as plt
import numpy as np

def sketch_distrib_2states (n,de=1,kT=None,xmax=1.1,figsize=(4,5)):

    # Makes a cartoon of occupancy plot.
    gridspec = {'width_ratios':[1,2*xmax]}
    fig, (ax1,ax2) = plt.subplots(1,2,figsize=figsize,
                                 gridspec_kw=gridspec)

    # The basic logic of the following sketch is the same
    # as sketch_occupancy() from an ealier chapter,
    # but it also includes an extra bar graph.

    # Define the size of boxes
    marg = 0.05 # Size of margin
    h = 1.0-2*marg
    w = 1.0-2*marg
    xbox = np.array([marg,marg+w,marg+w,marg])
    ybox = np.array([marg,marg,marg+h,marg+h])
    colors = ['#999999','#DDDDDD'] # darker, lighter gray
```

```python
    n = np.array(n)

    for i in range(2):
        ax1.fill(xbox,ybox+i,color=colors[i])
        x = (np.random.uniform(size=n[i])-0.5)*w*0.9+0.5
        y = (np.random.uniform(size=n[i])-0.5)*h*0.9+0.5+i
        ax1.scatter(x,y,marker='.',color='k',s=50,zorder=2.5)
        # Display the fraction to the left of each box.
        ax1.text(-0.35,i+0.2,'%3.2f'%(n[i]/np.sum(n)))
    ax1.set_ylim(0,2)
    ax1.set_yticks([])
    ax1.set_xticks([])
    ax1.set_aspect('equal')
    ax1.axis('off')

    # If kT is not specified as an optional input argument,
    # calculate equilibrium kT given n[0] and n[1].
    if kT==None:
        kT = -de/np.log(n[1]/n[0])

    f = n/np.sum(n) # fraction of occupancy.
    e = np.array([0,de])
    for i in range(2):
        ax2.bar(e[i],f[i],width=0.1,color=colors[i])

    Z = 1 + np.exp(-de/kT) # Partition function

    de_range = np.arange(0,xmax,0.01) # energy difference
    boltzmann_dist = np.exp(-de_range/kT)/Z
    ax2.plot(de_range,boltzmann_dist,color='black')
    ax2.set_xlabel('$\epsilon$')
    ax2.set_xticks(np.arange(0,xmax+0.1,0.5))
    ax2.set_xlim((-0.1,xmax+0.1))
    ax2.set_ylim((0,1))
    ax2.set_yticks((0,0.5,1))
    ax2.set_ylabel('Fraction')
    ax2.set_aspect('equal')
    plt.tight_layout()
    return kT

n = [40,10]
de = 0.5
sketch_distrib_2states(n,de=de,figsize=(4,5))
plt.savefig('fig_ch9_sketch_distrib_2states_demo.eps',
            bbox_inches='tight')
plt.show()
```

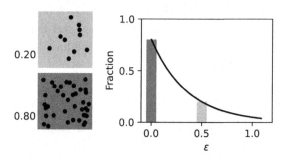

Figure 9.2

The bar graph carries additional information. The horizontal separation between the two bars corresponds to the energy level difference. In the above example, $\Delta\epsilon$ equals 0.5 (**de = 0.5**).

The exponential curve illustrates what the expected normalized Boltzmann distribution curve, $\frac{e^{-\Delta\epsilon/k_BT}}{Z}$ should be. Hence, at thermal equilibrium, the top of the two bars should touch the curve, as they do in Figure 9.2. Because $\frac{N_2}{N_1} = e^{-\Delta\epsilon/k_BT}$, the value of k_BT at thermal equilibrium can be obtained as $-\Delta\epsilon/\ln\left(\frac{N_2}{N_1}\right)$ for given values of N_1 and N_2, and this calculation shows up within the above code block as **kT = -de/np.log(n[1]/n[0])**.

Using this handy visualization function, let's create a series of plots that illustrates how a two-state thermal system behaves over a range of $\Delta\epsilon/k_BT$ values. For simplicity, we will fix k_BT at a constant value (**kT = 1** in the code). As the energy gap $\Delta\epsilon$ increases, we can explore how relative occupancy changes. As expected, for a given amount of average thermal energy of k_BT, the higher energy gap means that the distribution of particles will become more heavily skewed toward the lower energy level. As an analogy, when the price difference between the upper- and lower-level apartments increases, more people will take the cheaper, lower-level units. The following plots in Figure 9.3 show this trend by a higher density of dots and a higher fraction of particles in the lower energy level, as $\Delta\epsilon$ increases or as the separation between the bars increases.

The particle distributions, or the exact values of N_1 and N_2, were calculated as follows. Let $r = \frac{N_2}{N_1} = e^{-\Delta\epsilon/k_B T}$, as established earlier. Then, $N_2 = N_1 r$. Combining it with $N_2 = N - N_1$, we have $N_1 = N/(1+r)$ and $N_2 = Nr/(1+r)$.

```
# Code Block 9.3
# Sketch a series of distribution plots for a range of energy gaps.

N_total = 50
kT = 1
de_range = np.array([0.1,0.5,1,2])
for i, de in enumerate(de_range):
    r = np.exp(-de/kT)
    N1 = np.round(N_total/(1+r))
    N2 = np.round(N_total-N1)
    n = np.array([N1,N2],dtype='int')
    sketch_distrib_2states (n,de=de,kT=kT,xmax=2,figsize=(6,8))
    str = '$\Delta \epsilon/kT$ = %2.1f, $\Delta \epsilon$ = %2.1f'
    plt.title(str%(de/kT,de))
    plt.savefig('fig_ch9_occupancy_fixed_T_%d.eps'%i,
                bbox_inches='tight')
    plt.show()
```

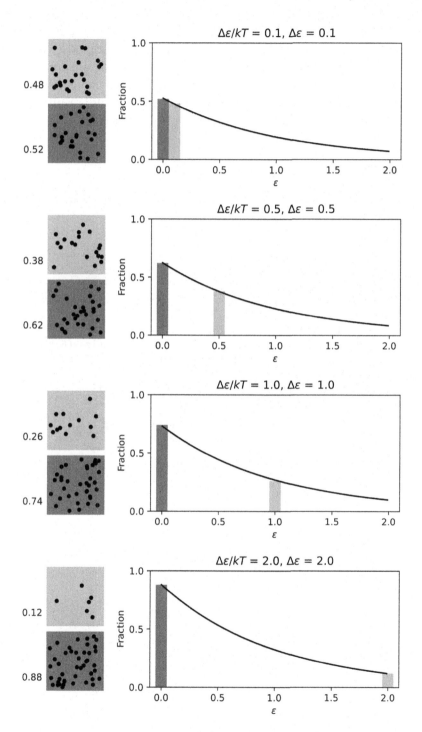

Figure 9.3

In the next set of visualizations, let's vary $k_B T$ while keeping $\Delta \epsilon$ fixed (**de** = **0.5** in the code). As the average thermal energy $k_B T$ increases, the energy gap between the two levels becomes less significant in determining the relative occupancy. Therefore, at higher values of $k_B T$, the occupancy of the two levels becomes more comparable, resulting in an almost equal number of particles at both energy levels.

```
# Code Block 9.4

# Sketch a series of distribution plots for different thermal energy.

N_total = 50
de = 0.5
kT_range = np.array([0.25,0.5,1,5])
for i, kT in enumerate(kT_range):
    r = np.exp(-de/kT)
    N1 = np.round(N_total/(1+r))
    N2 = np.round(N_total-N1)
    n = np.array([N1,N2],dtype='int')
    sketch_distrib_2states (n,de=de,kT=kT,xmax=2,figsize=(6,8))
    str = '$\Delta \epsilon/kT$ = %2.1f, kT = %3.2f'
    plt.title(str%(de/kT,kT))
    plt.savefig('fig_ch9_occupancy_fixed_de_%d.eps'%i,
                bbox_inches='tight')
    plt.show()
```

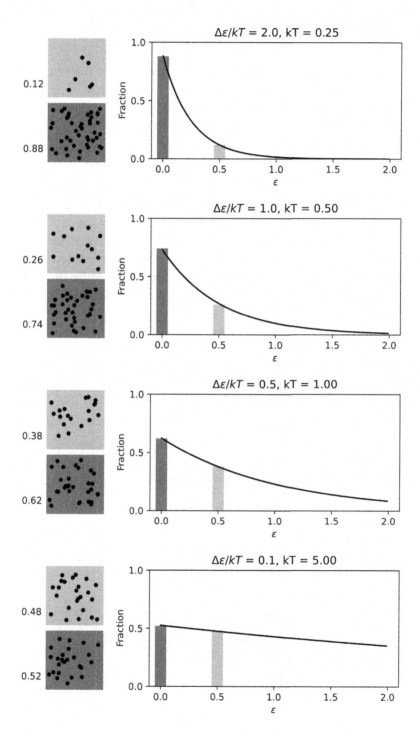

Figure 9.4

9.1 DYNAMIC CASE

Our analysis thus far has dealt with the occupancy of the energy levels at thermal equilibrium, which is determined by the ratio of $\Delta \epsilon$ and $k_B T$. What would happen when the system is not quite at thermal equilibrium? The answer to the question is that the system will move toward the state described by the Boltzmann distribution since this state has the most microstates or the highest entropy, and hence it is the most probable.

The next code block demonstrates a two-state system with a fixed energy difference, $\Delta \epsilon$ (de = 0.5 in the code) and constant $k_B T$. It is initially at a non-equilibrium state with many more particles in the lower energy level than expected from the Boltzmann distribution. This deviation from the Boltzmann distribution is represented by the fact that the top of the bars does not coincide with the exponential curve. However, this system at non-equilibrium moves closer to the steady state of the Boltzmann distribution, as some particles migrate from lower to upper energy levels. The extra energy that allows the promotion of these particles would come from the environment. Particles will randomly move back and forth between the energy levels by releasing or absorbing energy. Over time, the number of particles in each energy level will match the values expected from the Boltzmann distribution, and a steady state will have been reached.

```python
# Code Block 9.5
# Sketch of transition toward equilibrium state.

N_total = 50
de = 0.5
kT = 1
r_range = np.array([0.1,0.2,0.4,0.6])
for i, r in enumerate(r_range):
    N1 = np.round(N_total/(1+r))
    N2 = np.round(N_total-N1)
    n = np.array([N1,N2],dtype='int')
    sketch_distrib_2states (n,de=de,kT=kT,figsize=(4,5))
    plt.savefig('fig_ch9_occupancy_dynamic_%d.eps'%i,
                bbox_inches='tight')
    plt.show()
```

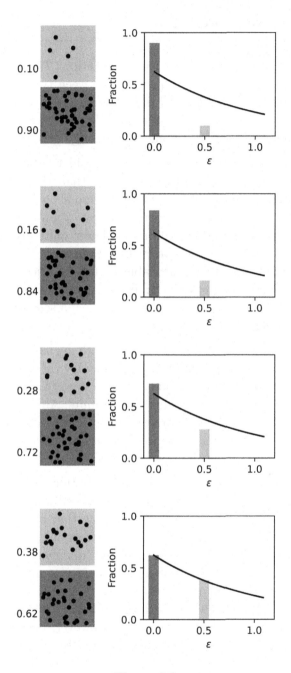

Figure 9.5

9.2 EQUILIBRIUM POTENTIAL

Let's consider an even more dynamic situation. The above situation assumed that the movement of particles between the energy levels did not affect the energy gap $\Delta\epsilon$, and the system was simply moving toward the Boltzmann distribution that is static. However, what would happen if the movement of particles changed the energy gap, too?

As a motivating analogy, we may again consider a two-story building where the upper units are more expensive than the lower units, but the price difference is no longer fixed. As more people move into the upper units, the price gap may increase due to the supply-and-demand effect. Suppose most people lived in the lower units initially. As some people start to move into the upper units, the supply of the upper units decreases, and hence, it is conceivable that their price might increase, while the price of the lower units drops. As a result, moving to an upper unit becomes financially more difficult. In such a scenario, would there be an equilibrium market price?

It is worth noting that there are a few underlying assumptions. For example, where would the extra wealth to pay for the upper-level apartments come from? We may assume that an upward move causes a collective drop in the price of the lower-level units. Alternatively, we may assume that the town is not a completely closed system, and there could be some fluctuation in the total wealth of the town due to a flow of money into and out of the town from the outside.

This analogy hints that the population distribution and the gap between the two levels can be closely related. A similar situation can be created by a semi-permeable divider or a membrane that can selectively pass a specific type of ion, such as Na^+. Suppose we fill a tank with water, place this special divider in the middle, and dissolve the salt, NaCl, on one side of the tank. NaCl will break up into Na^+ and Cl^- ions. Because the divider is permeable only to Na^+ ions, they would move across the membrane, mostly in the direction of equalizing the concentration of Na^+ between the two sides of the tank. However, because Cl^- cannot cross the divider, the movement of Na^+ ions will result in the differentiation of the net electrical charges across the divider. There will be a net increase of negative electrical charges on one side of the tank where NaCl was initially added since Cl^- ions will outnumber Na^+ ions. The other side will have a net increase of positive electrical charges due to the influx and net accumulation of Na^+ ions since Cl^- ions cannot cross the divider.

The like charges (e.g., two positive charges) repel each other, and the opposite charges (e.g., a positive and a negative) attract each other, according to Coulomb's well-known law of electromagnetism. Hence, the net increase of Na^+ ions on one side would not continue indefinitely. At some point, the repulsive electrical force between Na^+ ions will be balanced by the diffusive movement of Na^+ ions from the side of higher concentration. In other words, the voltage difference between the two sides would eventually be too high for a Na^+ ion to overcome despite the concentration difference.

The voltage difference is the amount of electrical potential energy per unit charge. Just as a mass at a certain height possesses gravitational potential energy, an electrical charge on an electrical landscape possesses electrical energy. An electrical landscape is created by a distribution of charges in space. Just as a higher-mass object possesses more gravitational energy than a lower-mass object at the same height, an object with a higher electrical charge has higher electrical potential energy than a lower-charge object at the same voltage difference.

It takes work (in the physics sense of applying force over distance) to bring a positive charge toward a crowd of other positive charges because the repulsive force has to be overcome. Therefore, a positive charge brought to a region of other positive charges has gained electrical potential energy. In other words, this region of positive charges has a positive electrical potential compared to the origin of the single charge. Unfortunately, we are dealing with two terms that are conceptually different but sound confusingly similar: Potential energy versus potential. Potential is associated with different positions, and potential energy is gained or lost by an object when it moves between places with different potentials. Electrical potential difference and voltage difference refer to the same concept and are measured in the unit called volts, which is equivalent to joules per coulomb, or a unit of energy divided by a unit of charge.

Let's calculate this voltage difference in terms of the ionic concentrations (or the number of ions) on the two sides of the tank. According to the Boltzmann distribution, the ratio of the number of ions is:

$$\frac{N_2}{N_1} = e^{-\Delta\epsilon/k_B T}.$$

The energy difference $\Delta\epsilon$ for each ion is equal to the product of its electrical charge and the potential difference V across the membrane. Hence, $\Delta\epsilon = -ZeV$, where $|Ze|$ is the amount of electric charge in Coulombs of each ion. For Na$^+$, $Z = 1$. Following this, the above expression can be simplified as

$$\ln\frac{N_2}{N_1} = \frac{ZeV}{k_B T}$$

or

$$V = \frac{k_B T}{Ze} \ln\frac{N_2}{N_1}.$$

This is called the Nernst equation. It captures a delicate balance between two competing tendencies within this special tank: (1) a diffusive, entropic movement of ions from a region of higher concentration to a lower region, and (2) an energetic movement of positive ions from higher to lower potential.

The following code block simulates this situation with a two-state system. We will start with most particles in the lower energy level and a small energy difference $\Delta\epsilon$. The temperature is fixed to a constant value ($kT = 1$ in the code). Let's make a simple assumption that the energy gap linearly increases with the migration of each particle from lower to upper energy level, or $\Delta\epsilon$ is proportional to the number of particles in the upper level:

$$\Delta\epsilon = \frac{1}{C}N_2,$$

where $\frac{1}{C}$ is the proportionality constant (C is conceptually similar to an electrical capacitance which is equal to Q/V).

At each point in time, $\Delta\epsilon/k_B T$ is determined, and there is an expected occupancy value, according to the Boltzmann distribution. We can make another simple assumption that the rate of particle movement across the energy levels is proportional to the difference between the expected and actual number of ions at each level. There will be more movements when the two-state system is very far from the equilibrium or the discrepancy between the expected and actual numbers is large. There will be fewer movements when the system is almost at equilibrium. In the code, this idea is implemented by `discrepancy = n1_actual - n1_expected` and `n[1] = int(n[1] + discrepancy*efficiency)`, where the variable **efficiency** determines how quickly and efficiently the ions move across the divider, given discrepancy.

```
# Code Block 9.6

# Sketch of transition toward an equilibrium state
# with varying energy gap.

kT = 1

C = 50 # capacitance or rate of change in energy gap per particle.
efficiency = 0.3 # efficiency of particle movement across membrane.

# initial condition.
n = np.array([95,5],dtype='int')

# Different initial conditions to try.
#n = np.array([50,50],dtype='int')

N_total = np.sum(n)
for i in range(8):
    de = n[1]/C # energy difference
    sketch_distrib_2states(n,de=de,kT=kT,figsize=(4,5))
    plt.savefig('fig_ch9_dynamic_2states_%d.eps'%i,
                bbox_inches='tight')
    plt.title('$\Delta \epsilon = %3.2f, kT = %2.1f$'%(de,kT))
    plt.show()

    # Update the occupancy.
    r = np.exp(-de/kT)
    n1_expected = np.sum(n)/(1+r)
    n1_actual = n[0]
    discrepancy = n1_actual - n1_expected
    n[1] = int(n[1] + discrepancy*efficiency)
    n[0] = N_total-n[1]
```

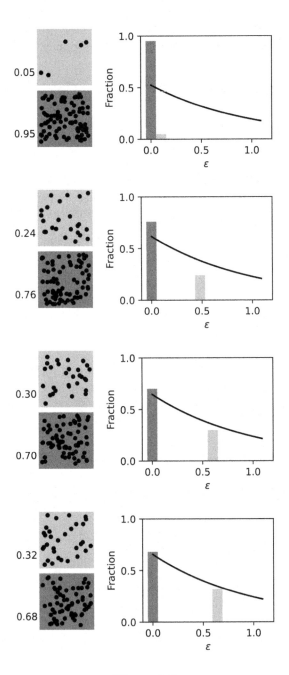

Figure 9.6

The above plots in Figure 9.6 show how the two-state system moves dynamically toward an equilibrium state over time. Note that the target equilibrium state, shown as an exponential curve, is also changing since $\Delta\epsilon$ changes with each movement of ions. There are two competing drives: entropic (due to the concentration gradient) and energetic (due to the voltage gradient), and eventually, an equilibrium state is reached, so that the actual relative occupancy matches the expected value. Also, we can see that there is an equilibrium value of $\Delta\epsilon$ for the given configuration of the thermal system (i.e., an equilibrium market price for an apartment unit is reached).

9.3 ACTION POTENTIAL

Such a selective passage of ions happens at a cellular membrane, which contains specialized proteins called ion channels. These proteins allow specific ions, such as Na^+ or K^+, to flow into or out of the cell body. What is impressive is that most ion channels are not just passive pores, but they would open and close under specific conditions. Some ion channels, known as voltage-gated, open up when the voltage difference across the cellular membrane crosses some threshold. Other ion channels react to mechanical pressure on the membrane, to the binding of a precisely-shaped molecule, or to the absorption of photons. Just as there are many different doors with specialized locking-unlocking mechanisms, there are numerous types of ion channels.

Another specialized protein on a cellular membrane is called an ion pump. It uses the energy, usually in the form of ATP, to actively transport ions against the concentration gradient. For example, a sodium-potassium pump uses the energy from one ATP molecule to export three Na^+ ions and import two K^+ ions, with a net export of a positive electrical charge per pumping cycle. As a result, there are a higher concentration of Na^+ ions outside and a lower concentration of K^+ ions inside of a cell body, and combined with the influences of other ions, the resulting electrical potential difference is 70 mV, where the intracellular potential is lower than the outside of the cell.

Each cell invests its energy to establish a concentration gradient and voltage difference, so that when the ion channels open up due to a specific stimulus, the resulting movement of ions causes a rapid fluctuation of the electric potential difference across the membrane, which is known as an action potential, and its propagation can be used as a signaling mechanism. Such a situation may be compared to the work of a skier who laboriously brings herself to the top of a mountain and then enjoys a rapid glide down the slope.

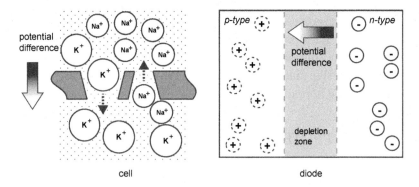

cell   diode

Figure 9.7

9.4   DIODE

A similar mechanism (involving diffusion of electrical charges and development of electrical potential) underlies the operation of many electronic devices. The semiconductor material can be given different properties through a manufacturing process called doping. In an n-type semiconductor, many free electrons serve as charge carriers of electric current. In the p-type, the absence of an electron, also known as a hole, serves as a positive charge carrier. When p-type and n-type semiconductors are put next to each other, the relative abundance of different charge carriers will drive the diffusion across the boundary between the two semiconductors. More electrons will migrate toward the p-type semiconductor, where they will fill the existing holes. The opposite happens in the n-type semiconductor, where the holes from the p-type semiconductor will diffuse into the n-type semiconductor.

The net result of the diffusion is that near the boundary of the two types of semiconductors, there will be a lack of charge carriers. This special layer is called the depletion region. Furthermore, near the boundary of the junction, the p-type side will have an overall negative charge due to the migrated electrons. Similarly, the n-type side will develop an overall positive charge near the boundary. In other words, an equilibrium potential (so-called built-in voltage, typically, around several hundreds of mV, depending on the semiconductor materials used) develops across the junction. An external voltage may be applied in a forward or reverse direction and control the flow of electric current. This arrangement of two semiconductor materials is called a diode or a p-n junction. More elaborate arrangements of semiconductors are possible and, of course, underlie modern technological advances in electronics.

## CHAPTER 10

Specific Heat

10.1 DEFINITION OF SPECIFIC HEAT

Specific heat is an amount of heat energy required to raise the temperature of a given substance, and its formal definition is:

$$C = \left(\frac{dQ}{dT}\right).$$

It is a quantity that can be measured through an experiment. For example, one may add a known quantity of heat (by burning a prescribed amount of fuel or by delivering a known amount of electrical power) and measure the change in the temperature of a substance in question.

The specific heat may differ depending on the exact process of how heat is added to the substance. For example, heat may be added while the volume or the pressure of the substance is held constant, as in our discussion of the thermal process of an ideal gas in Chapter 4.

The specific heat can be described in a more general way by treating the added heat energy with the thermodynamic potentials we have introduced earlier. For specific heat C_V at constant volume, let's consider the following relation:

$$dU(V, T) = \left(\frac{\partial U}{\partial V}\right)_T dV + \left(\frac{\partial U}{\partial T}\right)_V dT.$$

Since $dU = dQ - PdV$ with a fixed number of particles, these two expressions lead to

$$C_V = \left(\frac{dQ}{dT}\right)_V = \left(\frac{\partial U}{\partial T}\right)_V.$$

For specific heat C_P at constant pressure, let's consider enthalpy $H(P,T)$:

$$dH(P,T) = \left(\frac{\partial H}{\partial P}\right)_T dP + \left(\frac{\partial H}{\partial T}\right)_P dT.$$

Since $H = U + PV$, the differential form of enthalpy can be also written as $dH = dU + PdV + VdP = dQ + VdP$ where we used $dU = dQ - PdV$ again. By comparing two different forms of H, we conclude

$$C_P = \left(\frac{dQ}{dT}\right)_P = \left(\frac{\partial H}{\partial T}\right)_P.$$

As a quick check, we can apply these general definitions of C_V and C_P to the ideal gas case. The internal energy of ideal gas is given as $U = \frac{3}{2}Nk_BT$, and its enthalpy as $H = U + PV = \frac{3}{2}Nk_BT + Nk_BT = \frac{5}{2}Nk_BT$. Hence, $C_V = \frac{3}{2}Nk_B$ and $C_P = \frac{5}{2}Nk_B$ for ideal gas as expected from Mayer's equation.

This chapter will consider different thermal systems and examine their specific heats. We will start with a simple two-state system and a simple harmonic oscillator and extend our discussion to a case of a solid by modeling it as a collection of atoms. We will use the following result from Chapter 6:

$$U = Nk_BT^2\frac{\partial \ln Z}{\partial T},$$

which applies to any thermal system.

10.2 TWO-STATE SYSTEM

For a two-state system with an energy difference of $\Delta\epsilon$, we have determined that its partition function is

$$Z = 1 + e^{-\Delta\epsilon/k_BT}.$$

Therefore, the internal energy of the two-state system is

$$U = Nk_B T^2 \frac{\partial \ln Z}{\partial T}$$

$$= Nk_B T^2 \frac{\Delta \epsilon}{k_B T^2} e^{-\Delta \epsilon / k_B T} \frac{1}{Z}$$

$$= \frac{N}{Z} \Delta \epsilon e^{-\Delta \epsilon / k_B T}.$$

Then, C_V can be calculated by taking another derivative with respect to T. For simplicity, we will consider internal energy per particle, U/N and specific heat per particle, C_V/N with $\Delta \epsilon / k_B = 1$ in the following code block.

```
# Code Block 10.1

# Calculate internal energy and specific heat of a two-state system.

import sympy as sym
import matplotlib.pyplot as plt
import numpy as np

de = sym.Symbol('\Delta \epsilon')
k = sym.Symbol('k_B')
T = sym.Symbol('T')

Z = 1+sym.exp(-de/(k*T))
u = k*T**2*sym.diff(sym.ln(Z),T)
c = sym.diff(u,T)

print('Internal Energy, U/N')
display(u)
print(' ')
print('Specific Heat, C/N')
display(c)

# Calculate numerical values.
T_range = np.arange(0,3,0.01)
u_range = np.zeros(len(T_range))
c_range = np.zeros(len(T_range))
for i in range(len(T_range)):
    u_range[i] = u.subs({k:1,de:1,T:T_range[i]}).evalf()
    c_range[i] = c.subs({k:1,de:1,T:T_range[i]}).evalf()

plt.plot(T_range,u_range,color='#000000',linestyle='dotted')
plt.plot(T_range,c_range,color='#AAAAAA',linestyle='solid')
plt.legend(('Internal Energy, U/N','Specific Heat, C/N'),
           framealpha=1)
```

```
plt.xlabel('T')
plt.ylabel('U/N (Joules), C/N (Joules/Kelvin)')
plt.title('$\Delta \epsilon/k_B = 1$')
plt.savefig('fig_ch10_specific_heat_2states.eps')
plt.show()
```

Internal Energy, U/N

$$\frac{\Delta \epsilon e^{-\frac{\Delta \epsilon}{Tk_B}}}{1 + e^{-\frac{\Delta \epsilon}{Tk_B}}}$$

Specific Heat, C/N

$$\frac{\Delta \epsilon^2 e^{-\frac{\Delta \epsilon}{Tk_B}}}{T^2 k_B \left(1 + e^{-\frac{\Delta \epsilon}{Tk_B}}\right)} - \frac{\Delta \epsilon^2 e^{-\frac{2\Delta \epsilon}{Tk_B}}}{T^2 k_B \left(1 + e^{-\frac{\Delta \epsilon}{Tk_B}}\right)^2}$$

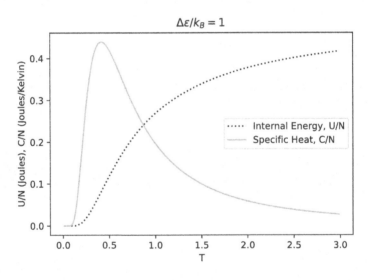

Figure 10.1

10.3 SIMPLE HARMONIC OSCILLATOR (SHO)

A canonical model of a simple harmonic oscillator (SHO) is a one-dimensional mechanical system with a lumped mass, like a ball or a block, attached to one end of a spring. When the mass is pulled or pushed against the spring, the spring provides the restoring force in the opposite direction, making the mass oscillate around the equilibrium position. The total mechanical energy of the simple harmonic oscillator is the sum of the kinetic energy of the moving mass and the potential energy stored in the spring. Even though the kinetic and potential energy at an arbitrary position might be different, the average value of each will be the same as half of the total energy.

A classical SHO has a continuous value of energy, but a quantum-mechanical SHO can be treated as a thermal system with an infinite number of energy levels with an equal energy difference $\Delta\epsilon$ between the adjacent levels. That is, a quantum SHO can take only one of the discrete energy levels, and the successive levels can be described as $n\Delta\epsilon$ where n is a non-negative integer (0, 1, 2, ...). The partition function for such a system can be written as[†]:

$$Z = 1 + e^{-\frac{\Delta\epsilon}{k_BT}} + e^{-\frac{2\Delta\epsilon}{k_BT}} + e^{-\frac{3\Delta\epsilon}{k_BT}} \cdots .$$

This is an infinite geometric series, converging to $Z = \frac{1}{1-e^{-\frac{\Delta\epsilon}{k_BT}}}$ since $e^{-\frac{\Delta\epsilon}{k_BT}} < 1$. We can obtain U and C_V for an SHO by following the same steps as for the two-level system.

```
# Code Block 10.2

# Calculate internal energy and specific heat of SHO.

import sympy as sym
import matplotlib.pyplot as plt
import numpy as np

de = sym.Symbol('\Delta \epsilon')
k = sym.Symbol('k_B')
T = sym.Symbol('T')

Z = 1/(1-sym.exp(-de/(k*T)))
u = k*T**2*sym.diff(sym.ln(Z),T)
c = sym.diff(u,T)

print('Internal Energy of SHO, U')
display(u)
print(' ')
print('Specific Heat of SHO, C')
```

[†]The lowest energy level, or the ground state, actually has a non-zero energy value, which we will consider later in the section about a solid. This ground state energy only introduces an overall additive constant for U and does not affect C_V.

```
display(c)

# Calculate numerical values.
T_range = np.arange(0,2,0.01)
u_range = np.zeros(len(T_range))
c_range = np.zeros(len(T_range))
for i in range(len(T_range)):
    u_range[i] = u.subs({k:1,de:1,T:T_range[i]}).evalf()
    c_range[i] = c.subs({k:1,de:1,T:T_range[i]}).evalf()

plt.plot(T_range,u_range,color='#000000',linestyle='dotted')
plt.plot(T_range,c_range,color='#AAAAAA',linestyle='solid')
plt.legend(('Internal Energy, U','Specific Heat, C'),framealpha=1)
plt.xlabel('T')
plt.ylabel('U (Joules), C (Joules/Kelvin)')
plt.title('$\Delta \epsilon/k_B = 1$')
plt.savefig('fig_ch10_specific_heat_SHO.eps')
plt.show()
```

Internal Energy of SHO, U

$$\frac{\Delta\epsilon e^{-\frac{\Delta\epsilon}{Tk_B}}}{1 - e^{-\frac{\Delta\epsilon}{Tk_B}}}$$

Specific Heat of SHO, C

$$\frac{\Delta\epsilon^2 e^{-\frac{\Delta\epsilon}{Tk_B}}}{T^2 k_B \left(1 - e^{-\frac{\Delta\epsilon}{Tk_B}}\right)} + \frac{\Delta\epsilon^2 e^{-\frac{2\Delta\epsilon}{Tk_B}}}{T^2 k_B \left(1 - e^{-\frac{\Delta\epsilon}{Tk_B}}\right)^2}$$

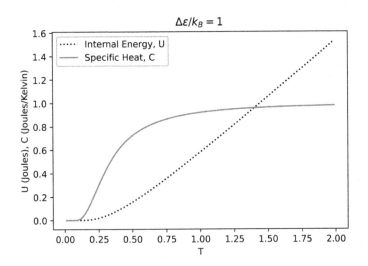

Figure 10.2

10.4 TEMPERATURE DEPENDENCE OF ENERGY AND SPECIFIC HEAT

For both thermal systems, most particles occupy the lowest energy level at a low temperature, so the thermal system has a low internal energy overall. As the temperature increases, the higher energy levels are increasingly populated. For the two-state system, when $k_B T >> \Delta \epsilon$, the occupancy of the higher energy level becomes comparable to the occupancy of the lower level, so the average internal energy $\frac{U}{N}$ is $\frac{1}{2}(0 + \Delta \epsilon)$. (The occupancy of the higher level can not be higher than the lower level, according to the Boltzmann distribution.) Therefore, the plot of U/N (dotted line) for the two-state system starts from zero at low T and approaches 0.5 at a higher temperature. (You can verify this by extending the range of temperature **T_range** in the corresponding code block. While doing so, to keep the computation time reasonable, set the temperature step larger.) There are infinitely many energy levels for an SHO system, so the internal energy can continue to increase with higher temperatures.

Specific heat (solid line) is zero for low temperatures because if $k_B T$ is not big enough to overcome the spacing $\Delta \epsilon$ between the energy levels, only a few particles will be able to jump to the higher energy levels. For a two-state system, there is an upper limit to the average internal energy, so the specific heat approaches zero again at a high temperature because further addition of heat can not make the energy higher than this upper limit. For an SHO system, as T increases, specific heat approaches a non-zero value, indicating that the system will continue to increase its internal energy as more heat energy is added. At high T, the available thermal energy $k_B T$ is much more significant than $\Delta \epsilon$, and a continuous variable may approximate the discrete energy levels. Then, the SHO may be compared to an ideal gas whose constituents can take on continuous values of kinetic energy. In the case of an ideal gas, we already noted that its average kinetic energy along one spatial dimension is equal to $\frac{1}{2}k_B T$, and hence $U = \frac{3}{2}N k_B T$ with three spatial dimensions. Similarly, a one-dimensional SHO will have an average kinetic energy of $\frac{1}{2}k_B T$, and it will also have an equal amount of average potential energy $\frac{1}{2}k_B T$. Such a result is called the equipartition theorem. Therefore, at high T, the internal energy of an SHO, the sum of kinetic and potential energies, is $\frac{1}{2}k_B T + \frac{1}{2}k_B T = k_B T$ and the specific heat would be k_B. In our code block, we scaled the constants, so that k_B (**k** in the code) is equal to 1. Hence, the limiting value of specific heat at high temperatures approaches 1 in the above plot.

10.5 EINSTEIN MODEL OF SOLID

We can expand on the SHO model and understand a solid's thermal behavior. Let us start with a case of two neighboring neutral atoms with potential energy as a function of the distance between them. There is an equilibrium separation where the potential energy is minimum with a net zero force between them. When two atoms get separated more than the equilibrium distance, a net attractive force brings the atoms closer. This attraction, commonly known as the van der Waals force, is due to the spontaneous formation of an electric dipole in an atom and the resulting dipole-dipole interaction. Its potential energy varies as the 6th power of the distance. Some other bonding mechanisms in a solid include: ionic bonds from the electrostatic interaction or covalent bonds through the sharing of electrons. The repulsive force pushes them apart when the two atoms get too close. This repulsion originates from the Pauli exclusion principle that bans the overlap of the electron clouds at a close range. The potential energy from the repulsive interaction varies as the 12th power of the distance. This potential energy model is called Lennard-Jones 6-12 potential. The combination of the attractive and repulsive interactions around an equilibrium point creates a potential well and is similar to the restoring force of a spring. Therefore, a three-dimensional solid composed of many atoms can be considered as a collection of simple harmonic oscillators that vibrate around their equilibrium positions.

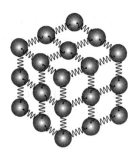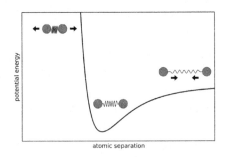

Figure 10.3

In the Einstein model of solid, each atom is regarded as an independent SHO with discrete energy levels. As in the one-dimensional SHO, the energy levels are equally spaced. Each energy level is described by the following:

$$\epsilon(n) = hf_E\left(\frac{1}{2} + n\right),$$

where h is Planck's constant and n is a non-negative integer. f_E is the characteristic oscillation frequency of an atom determined by the interactions with the neighboring atoms. It is analogous to the natural frequency of a classical oscillator determined by the ratio between spring constant and mass. Hence, the energy levels are equally spaced with $\Delta\epsilon = hf_E$. A key assumption in the Einstein model is that all atoms in the same solid oscillate with the same characteristic frequency f_E. Different materials with different atomic constituents would have their unique frequencies since their atom-to-atom interactions would differ.

The partition function for a single atom in a solid can be obtained by extending the approach for a one-dimensional SHO. Each atom in a three-dimensional solid can vibrate in x, y, and z-directions. Therefore, the complete partition function can be constructed as a product of three partition functions, each corresponding to a single direction:

$$Z = Z_x \times Z_y \times Z_z$$
$$= \left(e^{-\frac{hf_E}{2k_BT}} + e^{-\frac{3hf_E}{2k_BT}} + \cdots\right) \times \left(e^{-\frac{hf_E}{2k_BT}} + \cdots\right) \times \left(e^{-\frac{hf_E}{2k_BT}} + \cdots\right)$$
$$= \left(\frac{e^{-\frac{hf_E}{2k_BT}}}{1 - e^{-\frac{hf_E}{k_BT}}}\right) \times \left(\frac{e^{-\frac{hf_E}{2k_BT}}}{1 - e^{-\frac{hf_E}{k_BT}}}\right) \times \left(\frac{e^{-\frac{hf_E}{2k_BT}}}{1 - e^{-\frac{hf_E}{k_BT}}}\right)$$
$$= \left(\frac{e^{-\frac{hf_E}{2k_BT}}}{1 - e^{-\frac{hf_E}{k_BT}}}\right)^3$$

where we have used the fact that each series within parentheses is a converging series. The total internal energy of the solid with N atoms can be obtained with $U = Nk_BT^2\frac{\partial\ln Z}{\partial T}$. We will again use the **sympy** module in the following code block.

```
# Code Block 10.3

# Calculate internal energy and specific heat of Einstein solid.

import sympy as sym
import matplotlib.pyplot as plt
import numpy as np

h = sym.Symbol('h')
f = sym.Symbol('f_E')
k = sym.Symbol('k_B')
T = sym.Symbol('T')
N = sym.Symbol('N')

Z = (sym.exp(-h*f/(2*k*T))/(1-sym.exp(-h*f/(k*T))))**3
u = N*k*T**2*sym.diff(sym.ln(Z),T)
c = sym.diff(u,T)

print('Internal Energy of Einstein solid, U')
display(sym.factor(u))
print(' ')
print('Specific Heat of Einstein solid, C')
display(sym.factor(c))
print(' ')
print('Specific Heat at the high-temperature limit')

display(sym.limit(c,T,sym.oo))

# Calculate numerical values.
T_range = np.arange(0.01,2,0.01)
u_range = np.zeros(len(T_range))
c_range = np.zeros(len(T_range))
for i in range(len(T_range)):
    u_range[i] = u.subs({k:1,T:T_range[i],N:1,h:1,f:1}).evalf()
    c_range[i] = c.subs({k:1,T:T_range[i],N:1,h:1,f:1}).evalf()

plt.plot(T_range,u_range,color='#000000',linestyle='dotted')
plt.plot(T_range,c_range,color='#AAAAAA',linestyle='solid')
plt.legend(('Internal Energy, U','Specific Heat, C'),framealpha=1)
plt.xlabel('T')
plt.ylabel('U/N, C /N$k_B$')
plt.title('$h f_E/k_B = 1$')
plt.savefig('fig_ch10_specific_heat_einstein.eps')
plt.show()
```

Internal Energy of Einstein solid, U

$$\frac{3Nf_Eh\left(e^{\frac{f_Eh}{Tk_B}}+1\right)}{2\left(e^{\frac{f_Eh}{Tk_B}}-1\right)}$$

Specific Heat of Einstein solid, C

$$\frac{3Nf_E^2h^2e^{\frac{f_Eh}{Tk_B}}}{T^2k_B\left(e^{\frac{f_Eh}{Tk_B}}-1\right)^2}$$

Specific Heat at the high-temperature limit

$3Nk_B$

The above result of the symbolic calculation for specific heat can be written as:

$$C_{V, \text{ Einstein}} = 3Nk_B\frac{x^2e^x}{(e^x-1)^2},$$

where $x = \frac{hf_E}{k_BT}$.

The Einstein model of a solid is built upon the idea that each atom in a solid behaves like a simple harmonic oscillator, so the temperature dependence of its energy and specific heat can be understood in the same way as SHO. One difference from the one-dimensional SHO discussion is that the expression of the energy level includes a constant term, $\frac{1}{2}hf_E$. This is the lowest energy, also known as zero-point energy with $n = 0$, implying that even at an absolute zero temperature, the system still has non-zero energy. Therefore, at $T = 0$, the Einstein soild has non-zero internal energy. Since each atom in a three-dimensional solid has three degrees of freedom, the value of total zero-point energy is $3 \times \frac{1}{2}hf_EN$. Given our scaling of the constants in the code block (h, k, and f, corresponding to h, k_B, and f_E respectively, are all set to 1), U/N in Figure 10.4 approaches 1.5 at low T. At high temperatures, the N atoms in this three-dimensional solid collectively have a total energy of $3N$ times the energy of a one-dimensional SHO, which increases linearly with T. Therefore, the specific heat of an Einstein solid approaches $3Nk_B$, as shown in the above plot of C/Nk_B approaching 3. This trend is known as the Law of Dulong and Petit.

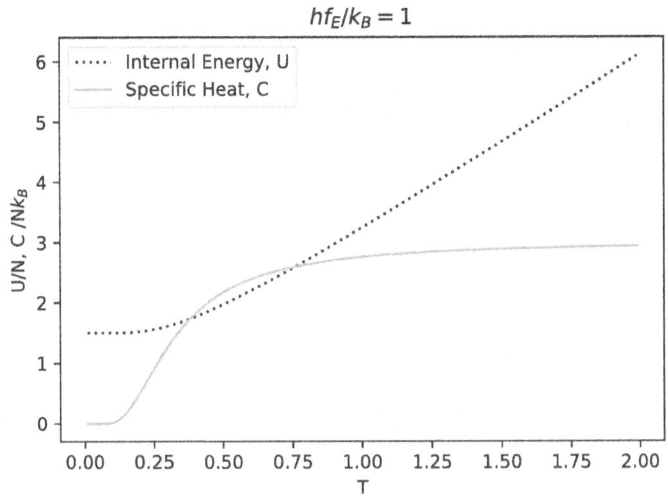

$hf_E/k_B = 1$

Figure 10.4

The Einstein solid is a rather crude model since it considers each atom as an independent oscillator with the same characteristic frequency. Although this simplistic assumption is good enough to capture the overall trend of C_V versus T, careful experimental measurements of C_V on bulk Al, Cu, and Ag show that there is a slight mismatch between the predictions of the Einstein model and the experimental data, especially at a lower temperature. Another model of a solid proposed by Debye provides a better fit to experimental data. The Debye model considers the collective motion of atoms with multiple frequencies, using the notion of phonons (similar to photons, but for the collective vibration responsible for the propagation of heat and sound in a solid). There is a maximum cut-off frequency known as the Debye frequency f_D. Debye's approach deals with the complex interactions among the coupled atoms more rigorously and matches the experimental data more accurately.

A fuller discussion of the Debye model will be left for other solid-state textbooks, and here we will simply present its result of specific heat:

$$C_{V,\,\text{Debye}} = 9Nk_B \left(\frac{k_B T}{hf_D}\right)^3 \int_0^{\frac{hf_D}{k_B T}} \frac{x^4 e^x}{(e^x - 1)^2} dx,$$

where x is a dimensionless integration variable. In the following code block, we will compare C_V's for the Einstein and Debye models. We utilized **np.sum()** function to perform the above integration numerically from zero up to the maximum Debye frequency f_D.

```python
# Code Block 10.4

# Comparing Einstein solid and Debye solid.

import numpy as np
import matplotlib.pyplot as plt

# Constants are set to a value of 1 for simplicity.
h = 1
k = 1
N = 1

T = np.arange(0.01,2,0.01)

# Integral necessary for calculating Debye model.
def Debye_integal(T_range,f_D=1,h=1,k=1):
    result = np.zeros(len(T_range))
    df = 0.01
    f_range = np.arange(df,f_D,df)
    for i, T in enumerate(T_range):
        dx = h*df/(k*T)
        x = (h*f_range)/(k*T)
        y = (x**4)*np.exp(x)/((np.exp(x)-1)**2) # integrand
        y = np.sum(y)*dx # numerical integration over frequencies.
        result[i] = y
    return result

# Debye solid.
f_D = 1
x = (h*f_D)/(k*T)
cD = 9*N*k*(x**(-3))*Debye_integal(T,f_D)

# Einstein solid
f_E = 1
x = (h*f_E)/(k*T)
cE = 3*N*k*(x**2)*np.exp(x)/(np.exp(x)-1)**2

plt.plot(T,cD,color='black',linestyle='solid')
plt.plot(T,cE,color='black',linestyle='dotted')
plt.legend(('Debye','Einstein'),framealpha=1)
plt.xlabel('T')
plt.ylabel('C / N$k_B$')
plt.title('Debye versus Einstein Models')
plt.savefig('fig_ch10_einstein_vs_debye.eps')
plt.show()
```

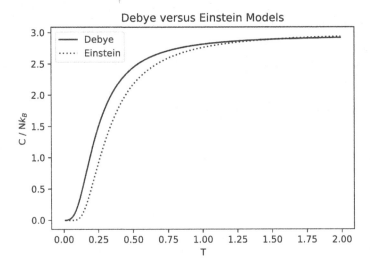

Figure 10.5

Specific heats of both Debye and Einstein solids at high temperatures approach the same constant value of $3Nk_B$, as expected from the Dulong and Petit law. However, the two solid models are not identical and display different temperature dependencies at low temperatures. The Debye solid model predicts that the drop of the specific heat is proportional to the cube of temperature, which is famously known as Debye's T^3 law, and it shows a better agreement with the experimental data.

## CHAPTER 11

Random and Guided Walks

11.1 ONE-DIMENSIONAL RANDOM WALK

During a random walk, a particle takes many random steps in succession and ends up at a position that may be far away from or close to its starting position. Such random movements, also known as a Brownian motion, are observed when a particle is immersed in a liquid or a gas and carries average kinetic energy in the order of $k_B T$, where T is the ambient temperature. Since no preferential direction is associated with a thermally-induced motion, the particle exhibits stochastic zigzag movements. In nanotechnology, such thermal fluctuations need to be understood and controlled precisely because they may pose serious challenges in developing a nano-scale mechanical system or be an exploitable source of ambient thermal energy. In finance, the rapid up and down swings in a stock market are explained using a similar framework as the mathematical model of a Brownian motion.

As a starting point, let's consider a random one-dimensional movement of a particle. After each successive time step, the particle moves a fixed distance, d, in either a positive or negative direction. Then, the total displacement $D(N)$ after N steps can be calculated as

$$D(N) = d_1 + d_2 + \cdots = \sum_{i=1}^{N} d_i,$$

where d_i is either positive or negative d.

If it is equally probable for the particle to take either positive or negative movements, then the average displacement $\mathbb{E}(D)$ will be zero. Therefore, the variance of $D(N)$ is $\mathbb{E}(D^2) - \mathbb{E}(D)^2 = \mathbb{E}(D^2)$. Both $\mathbb{E}(\cdot)$ and $< \cdot >$, which was introduced in an earlier chapter, are conventional notations for an average or a mean.

$$\mathbb{E}(D^2) = \mathbb{E}\left(\left(\sum_{i=1}^{N} d_i\right)^2\right) = \mathbb{E}\left(\left(\sum_{i=1}^{N} d_i\right)\left(\sum_{j=1}^{N} d_j\right)\right)$$

$$= \mathbb{E}\left(\sum_{i} d_i^2\right) + \mathbb{E}\left(2\sum_{i \neq j} d_i d_j\right).$$

Here, d_i^2 will always be equal to positive d^2. The last term $\mathbb{E}(d_i d_j) = 0$, because the movement at each time step is assumed to be independent and hence the number of times when the product $d_i d_j$ is $+d^2$ will be, on average, matched by the times when it is $-d^2$. Therefore, $\mathbb{E}(D^2) = Nd^2$. The key trend is that the variance increases linearly with N.

We simulate one-dimensional random walks of single and multiple particles in the following code blocks. A total of N steps are stored in an array **each_walk**, where each element is either -1 or 1. To cumulatively add up the displacement after each step, **np.cumsum()** is used.

```
# Code Block 11.1

# 1D Random walk for a single particle.

import numpy as np
import matplotlib.pyplot as plt

N = 1000
d = 1

# np.random.randint gives either 0 or 1,
# so (*2) and (-1) give either -1 or 1.
```

```
each_walk = np.random.randint(2,size=N)*2 - 1

disp = np.cumsum(d*each_walk)

plt.plot(disp,color='#AAAAAA')
plt.ylabel('Displacement')
plt.xlabel('Time Step')
plt.title('Single Random Walk')
plt.savefig('fig_ch11_random_walk_single.eps')
plt.show()
```

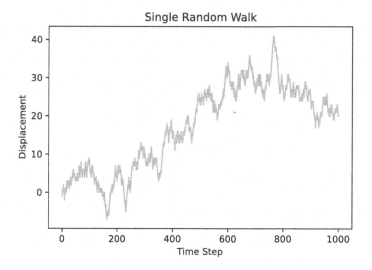

Figure 11.1

```
# Code Block 11.2

# We can simulate 1D random walks for multiple particles.

N_particles = 20
N = 1000
d = 1

# np.random.randint gives either 0 or 1,
# so (*2) and (-1) give either -1 or 1.
each_walk = np.random.randint(2,size=(N,N_particles))*2 - 1

disp = np.cumsum(d*each_walk,axis=0)
max_disp = np.abs(np.max(disp))
```

```
plt.plot(disp,color='#AAAAAA')
plt.ylim(np.array([-1,1])*1.5*max_disp)
plt.ylabel('Displacement')
plt.xlabel('Time Step')
plt.title('Multiple Random Walks')
plt.savefig('fig_ch11_random_walks.eps')
plt.show()

step = 25
plt.errorbar(range(0,N,step), np.mean(disp,axis=1)[::step],
            np.std(disp,axis=1)[::step], color='black')
plt.ylim(np.array([-1,1])*1.5*max_disp)
plt.xlabel('Time Step')
plt.ylabel('Displacement')
plt.title('Mean and Standard deviation of Random Walks')
plt.savefig('fig_ch11_random_walks_mean_std.eps')
plt.show()
```

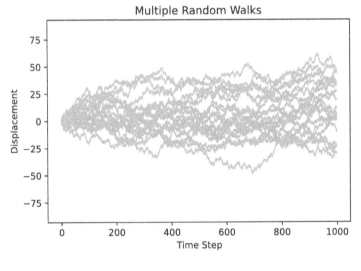

Figure 11.2

In the following code block, we show that $\mathbb{E}(D^2) = Nd^2$. By calculating the variance of multiple random walk trajectories using **np.var()** and plotting it against N, we can verify that $\mathbb{E}(D^2)/d^2$ versus N indeed gives a linear trend with a slope of 1.

```
# Code Block 11.3

# We can numerically confirm the linear trend of the variance.
# var(displacement) = N*(d^2), so the following plot should be
# close to a diagonal line (with some variations).

# We want to simulate many random walks, so
# increase N and N_particles.
N_particles = 2000
N = 100000

each_walk = np.random.randint(2,size=(N,N_particles))*2 - 1
disp = np.cumsum(d*each_walk,axis=0)

var_disp = np.var(disp,axis=1)

plt.plot(var_disp/(d**2),color='k')

# Draw a straight line for comparison.
plt.plot((0,N),(0,N),color='#AAAAAA')
plt.xlabel('Time Step')
plt.ylabel('Variance')
plt.axis('square')
plt.savefig('fig_ch11_var_time_linear.eps')
plt.show()
```

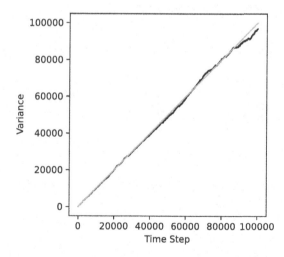

Figure 11.3

11.2 TWO-DIMENSIONAL RANDOM WALK

The above simulation of one-dimensional random walks can be easily extended to two-dimensional walks.

```python
# Code Block 11.4

# We can simulate 2D random walks for multiple particles.

N_particles = 6
N = 2000
d = 1
# np.random.randint gives either 0 or 1,
# so (*2) and (-1) give either -1 or 1.
each_walk_x = np.random.randint(2,size=(N,N_particles))*2 - 1
each_walk_y = np.random.randint(2,size=(N,N_particles))*2 - 1
disp_x = np.cumsum(d*each_walk_x,axis=0)
disp_y = np.cumsum(d*each_walk_y,axis=0)

for i in range(N_particles):
    # x0, y0 = initial position of a particle.
    # Stagger the locations so that they are easily distinguished.
    x0 = i*100
    y0 = 100*(-1)**i
    plt.plot(disp_x[:,i]+x0,disp_y[:,i]+y0,color='black')
plt.xlabel('Displacement in x')
plt.ylabel('Displacement in y')
plt.axis('equal')
plt.savefig('fig_ch11_random_walks_2dim.eps')
plt.show()
```

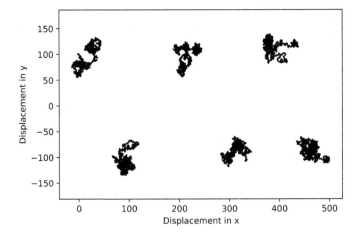

Figure 11.4

In the above simulation, only horizontal and vertical steps of size d were allowed, resulting in purely diagonal movements with an effective step size of $\sqrt{d}$. We can generalize the simulation with more possible directions, so that each step along x and y can be given by $(\Delta x, \Delta y) = (d\cos\theta, d\sin\theta)$, where θ can take on any value between 0 and 2π. It can be implemented in different ways. For example, we could randomly choose a direction with something like `np.random.rand()*2*np.pi`. Alternatively, we could divide the full range of directions evenly with something like `np.linspace(0,2*np.pi,num=num_dir,endpoint=False)` and choose one of those directions randomly. We implemented the latter approach within a function in the following code block.

```
# Code Block 11.5

# We can simulate 2D random walks for multiple particles
# along the random directions.

import numpy as np
import matplotlib.pyplot as plt

def random_walk_2D (N=2000,num_dir=10,d=1):
    possible_dir = np.linspace(0,2*np.pi,num=num_dir,endpoint=False)
    random_int = np.random.randint(num_dir,size=(N))
    theta = np.array([possible_dir[i] for i in random_int])
    x = np.cumsum(d*np.cos(theta))
    y = np.cumsum(d*np.sin(theta))
    return x,y

N_particles = 6
for i in range(N_particles):
    x,y = random_walk_2D()
    # x0, y0 = initial position of a particle.
    # Stagger the locations so that they are easily distinguished.
    x0 = i*100
    y0 = 100*(-1)**i
    plt.plot(x+x0,y+y0,color='black')
plt.xlabel('Displacement in x')
plt.ylabel('Displacement in y')
plt.axis('equal')
plt.savefig('fig_ch11_random_walks_2dim_general.eps')
plt.show()
```

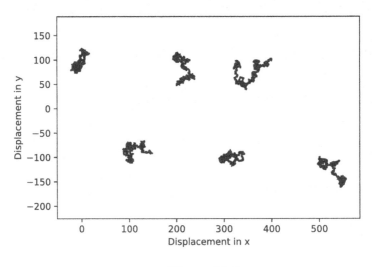

Figure 11.5

It is insightful to think about the trajectory of a random-walking particle as a long chain with free joints. The step size **d** in the simulation would be equivalent to the length of each unit of the chain, and the total number of walks **N** would be equivalent to the number of units that make up the chain. Hence, the maximum distance that can be traveled by a particle or the full length of the chain would be equal to **N*d**. A freely jointed chain means that the joint between two neighboring units of the chain can be at any angle, just as each step of the random walk can be in any direction. There is no energetic difference between a straight chain and a winding one.

The above simulations demonstrate that it is highly unlikely that a freely-jointed chain would be straight, just as a random walk is unlikely to produce a straight trajectory. Hence, a coiled or squeezed chain that looks like the above result is more entropically favored, even without any energetic bias. Such behavior is called entropic elasticity and is observed in some polymers that exhibit temperature-dependent elastic forces. One can do a simple experiment by hanging a rubber band, which is a polymer, vertically with a weight at the lower end. When the rubber band is heated (for example, with a hair dryer), the rubber band shrinks and lifts the weight. This shrinking is not necessarily dramatic since the weight would be lifted only slightly. Nevertheless, this result is a good illustration of how an entropic effect may be manifested mechanically. To many people, this result is surprising since thermal expansion is often more pronounced than shrinkage, as seen here.

More formally, it is possible to analyze this experiment with one of the thermodynamic potentials, F. As seen in Chapter 8, $F = U - TS$, and for a one-dimensional stretching and shrinking, we could use a linear length of a chain x as a state variable instead of the usual V. Then,

$$\frac{\partial F}{\partial x} = \frac{\partial U}{\partial x} - T\frac{\partial S}{\partial x}.$$

Since T does not depend on the length of the rubber band, there is no term with $\frac{\partial T}{\partial x}$. Assuming that the rubber band is a freely jointed chain, its internal energy U does not depend on its length, so $\frac{\partial U}{\partial x} = 0$. Hence,

$$f = -\frac{\partial F}{\partial x} = T\frac{\partial S}{\partial x},$$

where we are using one of the results from Chapter 8, $P = -\left(\frac{\partial F}{\partial V}\right)_{T,N}$. Instead of P, we use its linear analog f, which is the tension on the rubber band. The magnitude of tension increases with increasing T. The stretching of the rubber band decreases its entropy because a straight chain is unlikely, so $\frac{\partial S}{\partial x} < 0$. As a result, the overall sign of f becomes negative, implying that the tension f is restorative like a spring force. Hence, the shrinking of the rubber band when it gets hotter is explained as an entropic phenomenon.

11.3 A TANGENT

The above code can be adapted to create an interesting pattern out of rational and irrational numbers.* For example, we can use the successive digits of a number, such as 3141592... of π, as a direction at each step.

---

*This fun tangent was inspired by a Numberphile episode, "Plotting pi," at www.youtube.com/watch?v=tkC1HHuuk7c.

```
# Code Block 11.6

# Make a visual pattern of a number: pi or 1/7 or e.

import numpy as np
import matplotlib.pyplot as plt
import math
import mpmath as m # Multi-precision math
m.mp.dps = 20000

def guided_walk_2D (guide,d=1):
    possible_dir = np.linspace(0,2*np.pi,num=10,endpoint=False)
    theta = np.array([possible_dir[i] for i in guide])
    # The cumsum() makes the next heading angle be
    # relative to the previous direction.
    theta = np.cumsum(theta)
    x = np.cumsum(d*np.cos(theta))
    y = np.cumsum(d*np.sin(theta))
    return x,y

# Try different numbers by changing the case.
case = 'pi'
if case == 'pi':
    num = (4 * m.atan(1))/10
elif case == '1_over_7':
    num = m.power(7,-1)
elif case == 'e':
    num = m.exp(1)/10

num_str = m.nstr(num,n=m.mp.dps)[2:] # Ignore "0."

print('Digits of %s'%case)
print(num_str[:60])

# Convert individual digits into an array.
num = [int(x) for x in num_str]
x,y = guided_walk_2D(num)

# plot() function will work fine, but quiver() will also work.
#plt.plot(x,y,color='black',linewidth=0.25)
plt.quiver(x[:-1],y[:-1],x[1:]-x[:-1],y[1:]-y[:-1],
           units='xy',scale=1,color='black')
plt.axis('equal')
plt.axis('square')
plt.axis('off')
plt.savefig('fig_ch11_guided_walk_%s.eps'%(case),
            bbox_inches='tight')
plt.show()
```

Digits of pi
3141592653589793238462643383279502884197169399375105820974949

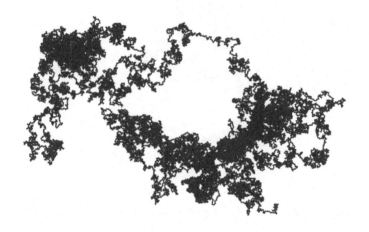

Figure 11.6

A fun mathematical musing is, if we let the simulation run infinitely long, would this plot trace all possible shapes (even a line sketch of a famous painting like Mona Lisa), since π never repeats. Other fascinating patterns can be generated with the repeating decimals of e or $1/7$ as shown below.

Figure 11.7

Figure 11.8

11.4 GUIDED RANDOM WALKS

We can simulate a target-seeking behavior by letting a particle's trajectory follow the gradient of the landscape. A gradient is a vector that quantifies the direction and magnitude of the steepest slope at a point. Formally, a two-dimensional gradient for a function $f(x, y)$ is given by:

$$\nabla f = \frac{\partial f}{\partial x}\hat{x} + \frac{\partial f}{\partial y}\hat{y},$$

where $\hat{x}$ and $\hat{y}$ are unit vectors in x and y directions. If you allow a particle to follow the gradient ∇f in small steps, the particle will keep climbing the landscape defined by $f(x, y)$. If you allow a particle to move in the opposite direction of the gradient in small steps, it will climb down the landscape.

This is precisely how a ball rolls downhill, following the negative gradient of gravitational potential energy and going toward a valley. This is how a positive charge moves along the negative gradient of electrical potential energy. In mechanics, the force on an object is the negative gradient of the potential energy. Therefore, the object moves to minimize its potential energy.

The following code block implements the above idea of gradient descent on a simple, parabolic function that has a global minimum at $(x_o, y_o) = (2, 3)$:

$$f(x, y) = (x - 2)^2 + 3(y - 3)^2.$$

Its gradient is:

$$\nabla f(x, y) = 2(x - 2)\hat{x} + 6(y - 3)\hat{y}.$$

Then, we can introduce a position update rule:

$$x(t + \Delta t) = x(t) + \Delta x = x(t) - d\left(\frac{\partial f}{\partial x}\right)$$

$$y(t + \Delta t) = y(t) + \Delta y = y(t) - d\left(\frac{\partial f}{\partial y}\right),$$

where d is a small step size. The negative sign in $\Delta x = -d\frac{\partial f}{\partial x}$ and $\Delta y = -d\frac{\partial f}{\partial y}$ indicates that the particle is descending, not ascending, along the function toward its minimum. After each update, the position becomes closer to the minimum of $f(x, y)$. The caveats are: if d is too large, the next position may overshoot the minimum, and if d is too small, many position updates are necessary to reach the minimum.

```python
# Code Block 11.7

# Simulating the gradient descent.

import numpy as np
import matplotlib.pyplot as plt

def gradient_simple_2D (x,y):
    # Calculates the gradient of a simple function.
    # f = (x-x0)**2 + 3*(y-y0)**2
    x0 = 2
    y0 = 3
    df_x = 2*(x-x0)
    df_y = 6*(y-y0)

    # Add noise.
    noise_level = 5
    df_x = df_x + np.random.randn()*noise_level
    df_y = df_y + np.random.randn()*noise_level
    return df_x,df_y

N = 200
d = 0.01 # step size.

x = np.zeros(N)
y = np.zeros(N)

# starting position
x[0] = -3
y[0] = -3
for i in range(N-1):
    dx, dy = gradient_simple_2D(x[i],y[i])
    x[i+1] = x[i] - d*dx
    y[i+1] = y[i] - d*dy

plt.plot(x,y,color='gray')
plt.plot(x[0],y[0],'ko')
plt.plot(x[-1],y[-1],'k*')
plt.text(x[0]+0.5,y[0],'start')
plt.text(x[-1]+0.5,y[-1],'end')

plt.axis('square')
plt.xlim((-5,5))
plt.ylim((-5,5))
plt.savefig('fig_ch11_guided_walk_simple.eps')
plt.show()
```

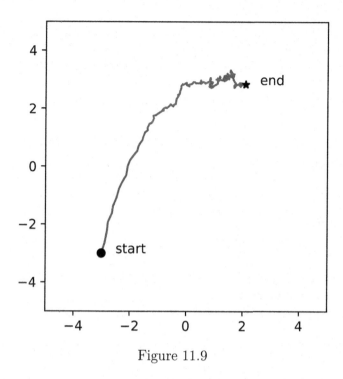

Figure 11.9

As shown in Figure 11.9, the particle started from an initial position of
x[0] and y[0]. It ended up near the global minimum of the landscape,
defined by the parabola with the vertex (x0, y0) = (2,3). There are
a few parameters in the above code block that can be experimented
with. For example, d determines how far the particle would travel at
each step. If d is too small, the particle would need more time (higher
N) to reach the target. If d is too large, the particle may not reach its
target position within a given time. The convergence may be improved
by starting with a large step size (for a fast, initial approach) and then
gradually decreasing it around the neighborhood of the target position
(for a more accurate convergence).

We also introduced a small noise by adding
np.random.randn()*noise_level to the gradient, so that a particle
would wander around a bit at each step, and it is another parameter
that can be experimented with. A different function $f(x, y)$ and its
gradient can be tried, too, by modifying gradient_simple_2D()
function.

The next code block extends the above by introducing multiple particles at different starting points. The collective target-seeking behavior is reminiscent of a population of ants converging on a food source. The concept of chemical potential μ from Chapter 8, which determines the movement of particles from one thermal system to another, is related to this idea.

```python
# Code Block 11.8

# Simulating the gradient descent for multiple particles.

P = 100 # number of particles
x = np.zeros((P,N))
y = np.zeros((P,N))

# starting position (randomly distributed)
x[:,0] = np.random.rand(P)*10-5
y[:,0] = np.random.rand(P)*10-5

# Calculate the trajectories.
for i in range(N-1):
    dx, dy = gradient_simple_2D(x[:,i],y[:,i])
    x[:,i+1] = x[:,i] - d*dx
    y[:,i+1] = y[:,i] - d*dy

# Display the trajectories of the particles.
for i in range(P):
    plt.plot(x[i,:],y[i,:],color='gray',linewidth=0.25)
    # Show the starting point of each particle.
    plt.plot(x[i,0],y[i,0],color='gray',marker='.')

# Show the average ending point.
plt.plot(np.mean(x[:,-1]),np.mean(y[:,-1]),'*',color='black')
plt.axis('square')
plt.xlim((-6,6))
plt.ylim((-6,6))
plt.savefig('fig_ch11_guided_walk_simple_multi.eps')
plt.show()
```

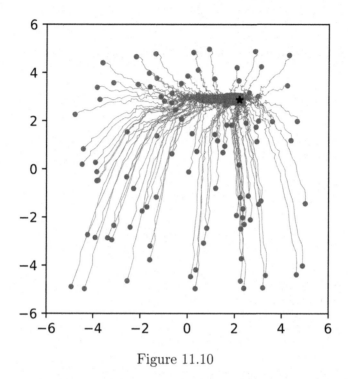

Figure 11.10

Next, consider a more complex landscape with two local minima, some-times called a double well. As a definite example, we will use

$$f(x,y) = -e^{-\left((x-x_0)^2+(y-y_0)^2\right)/4} - 2e^{-\left((x-x_1)^2+(y-y_1)^2\right)},$$

where $(x_0, y_0) = (-3, 3)$ and $(x_1, y_1) = (3, 3)$. The partial derivatives of the above function, as appear in `gradient_double_2D()` are

$$\frac{\partial f}{\partial x} = \frac{1}{2}(x - x_0)e^{-\left((x-x_0)^2+(y-y_0)^2\right)/4} + 4(x - x_1)e^{-\left((x-x_1)^2+(y-y_1)^2\right)}$$

$$\frac{\partial f}{\partial y} = \frac{1}{2}(y - y_0)e^{-\left((x-x_0)^2+(y-y_0)^2\right)/4} + 4(y - y_1)e^{-\left((x-x_1)^2+(y-y_1)^2\right)}$$

A one-dimensional slice through this two-dimensional function at $y = 3$ is given in the plot below and shows two local minima at x_0 and x_1. The global minimum is located at x_1.

```
# Code Block 11.9

# Simulating the gradient descent for double well.

def double_well_2D(x,y):
    x0, y0 = -3, 3
    x1, y1 =  3, 3
    f0 = -np.exp(-((x-x0)**2+(y-y0)**2)/4)
    f1 = -2*np.exp(-((x-x1)**2+(y-y1)**2))
    f = f0 + f1

    # Gradient
    df_x = -2*(x-x0)/4*f0 -2*(x-x1)*f1
    df_y = -2*(y-y0)/4*f0 -2*(y-y1)*f1

    return f, df_x, df_y

N = 100
d = 0.1 # step size
P = 100 # number of particles
x = np.zeros((P,N))
y = np.zeros((P,N))

# starting position (randomly distributed)
x[:,0] = np.random.rand(P)*10-5
y[:,0] = np.random.rand(P)*10-5

# Calculate the trajectories.
for i in range(N-1):
    _, df_x, df_y = double_well_2D(x[:,i],y[:,i])

    # Here is a slight modification.
    # We will make the gradient to be a unit vector,
    # so that each step size is always d.
    grad_norm = np.sqrt(df_x**2+df_y**2)
    df_x = df_x / grad_norm
    df_y = df_y / grad_norm

    x[:,i+1] = x[:,i] - d*df_x
    y[:,i+1] = y[:,i] - d*df_y

# Display the slice of the function, f(x,y)
x_range = np.arange(-5,5,0.01)
f, _, _ = double_well_2D(x_range,3)
plt.plot(x_range,f,color='k')

# Display where all particles ended on the slice.
x_range = x[:,-1]
```

```
f, _, _ = double_well_2D(x_range,3)

plt.plot(x_range,f,'ko')
plt.legend(('$f(x,y=3) = -e^{-(x+3)^2/4} -2e^{-(x-3)^2}$',
            'End points of trajectories'), framealpha=1.0)
plt.title('1D slice')
plt.xlabel('$x$')
plt.savefig('fig_ch11_guided_walk_double_slice.eps')
plt.show()

# Display the trajectories of the particles.
for i in range(P):
    plt.plot(x[i,:],y[i,:],color='gray',linewidth=0.25)
    # Show the starting point of each particle.
    plt.plot(x[i,0],y[i,0],color='gray',marker='.')
plt.axis('square')
plt.xlim((-6,6))
plt.ylim((-6,6))
plt.savefig('fig_ch11_guided_walk_double_multi.eps')
plt.show()

def find_neighbors (x,y,x0,y0,d_threshold):
    dist = np.sqrt((x-x0)**2 + (y-y0)**2)
    return np.sum(dist<d_threshold)

N0 = find_neighbors (x[:,-1],y[:,-1],-3,3,d*5)
N1 = find_neighbors (x[:,-1],y[:,-1], 3,3,d*5)
print('Number of neighbors near (%2d,%2d) = %d'%(-3,3,N0))
print('Number of neighbors near (%2d,%2d) = %d'%( 3,3,N1))
```

```
Number of neighbors near (-3, 3) = 71
Number of neighbors near ( 3, 3) = 29
```

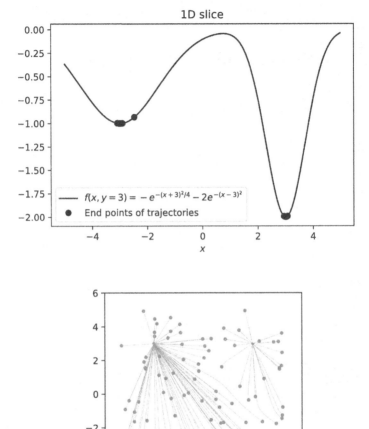

Figure 11.11

As illustrated in Figure 11.11 again, we can generate trajectories of multiple particles, which, as a result of following the negative of a gradient in small steps, discover local minima. The endpoints of the trajectories are shown as dots on the top one-dimensional slice plot, and the full trajectories are displayed on the bottom two-dimensional plot, where each dot represents a random starting point. In addition to calculating the trajectories on the landscape defined by $f(x, y)$, this code block also calculates the number of particles that arrived within the neighborhood of a specified point at the end of the walks. As expected, most particles end up near the vicinity of local minima. **d_threshold** parameter in the **find_neighbors()** function defines the range of this vicinity and determines whether a particle is included in the particle count or not.

This approach of discovering optimal points by guided walks underlies many successful optimization algorithms. For example, the back-propagation algorithm for an artificial neural network minimizes an error function by performing gradient descent within a high-dimensional phase space that represents the synaptic weights between neural units. Such an algorithm is necessary and important when the underlying function $f(\cdot)$ is not accessible or too complex, yet an estimate of its gradient $\nabla f(\cdot)$ is still available.

There are other interesting issues and challenges. For example, in the above simulation, more particles ended up near a local minimum at $x = -3$, even though the global minimum was not far away at $x = 3$. Can we find a global minimum and avoid being trapped in local minima? Can we obtain an accurate estimate of gradient, especially in a high dimensional space? Can we optimize only within relevant dimensions, without wasting our search efforts in unnecessary space? Can we build specialized hardware, perhaps taking advantage of quantum mechanics, for solving an optimization problem? These are some of the active research questions explored in such fields as statistics, optimization, machine learning, and physics.

Appendix

APPENDIX A: GETTING STARTED WITH PYTHON

Perhaps the most challenging step in following the codes in this book may be the first step of getting started with Python. Fortunately, there are a few user-friendly options at this point of writing.

The first option is a free, cloud-based Python environment like Google Colaboratory (or Colab) (**research.google.com/colaboratory**). You can open, edit, and run Python codes on a Jupyter Notebook environment using a browser. The second option is to download and install a distribution of Python that already includes relevant packages, such as **numpy** and **matplotlib**, and other valuable tools, such as Jupyter Notebook (**jupyter.org**). We recommend Anaconda Distribution (**www.anaconda.com**), which supports different operating systems (Windows, iOS, and Linux) and makes it easy to configure your computer.[†] The third option is to install each module and dependency separately.

APPENDIX B: PYTHON PROGRAMMING BASICS

Whether it is Python or other programming languages, there are many standard practices, notations, structures, and techniques. This appendix goes over a few basic ideas if you are new to programming.

The following code block demonstrates the practice of using variables to hold values and do calculations.

[†]There is an interesting interview of Travis Oliphant by Lex Fridman, where they talk about the history behind the development of **numpy**, **scipy**, Anaconda, and other topics on scientific computing available at **www.youtube.com/watch?v=gFEE3w7F0ww**.

```
# Code Block Appendix B.1

x = 5
y = 2
print(x+y)
print(x-y)
print(x*y)
print(x/y)
print(x**y)
```

```
7
3
10
2.5
25
```

Two powerful control structures are the **if**-conditionals and **for**-loops, as demonstrated below. The **for**-loop allows you to iterate a block of codes marked by indentation within the loop. The number of iterations is often specified with a built-in function **range()**. You can also easily work with individual elements in an array. If a condition given in the **if**-conditional is evaluated to be true, a set of codes marked by indentation will be executed.

```
# Code Block Appendix B.2

for i in range(5):
    print(i**2)
    if (i**2 == 9):
        print('This was a nice number.')
```

```
0
1
4
9
This was a nice number.
16
```

Another powerful practice in programming is to split up a complex task or procedure into smaller and more manageable chunks, which are called functions or modules. For example, you may be tasked to calculate an average of multiple values repeatedly. Then, it would be desirable to create a function that takes an arbitrary array of values as an input argument and returns its average.

In addition to being able to write your own functions or modules, it is also essential to be able to use well-written and widely-adopted modules. For example, many common and critical computational routines are already written into the modules like **numpy** and `scipy`. By using them, instead of writing your own, your codes will be more readable and usable by others and will likely be more robust and less susceptible to errors.

```
# Code Block Appendix B.3

import numpy as np

def calculate_average(x):
    avg = 0
    for val in x:
        avg = avg + val
    return avg/len(x)

x = [1,5,3,7,2]

# Using a function created above.
print(calculate_average(x))

# Using a function from numpy module.
print(np.mean(np.array(x)))
```

3.6
3.6

Another important aspect of coding is to make mistakes and learn from them. The following code blocks demonstrate a few common error messages you might see.

```
# Code Block Appendix B.4

# Using numpy module without importing it ahead will
# generate an error message like:
# "NameError: name 'np' is not defined"

# Because we are demonstrating the importance of import,
# let's unimport or del numpy.
del numpy

x = np.array([1,2,3])
```

```
---------------------------------------------------------------
NameError                         Traceback (most recent call last)
<ipython-input-4-976cd16bfa8a> in <module>
      7 # Because we are demonstrating the importance of import,
      8 # let's unimport or del numpy.
----> 9 del numpy
     10
     11 x = np.array([1,2,3])

NameError: name 'numpy' is not defined
```

```
# Code Block Appendix B.5

# Python's indexing convention is to start at zero.
# The first element in an array is indexed by 0.
# The last element in an array is indexed by -1 or its length-1.
# If you try to index after the last element,
# you will get an error like:
# "IndexError: list index out of range"

x = [10,20,30]
print(x[0])
print(x[1])
print(x[2])
print(x[3])
```

```
10
20
30
---------------------------------------------------------------
IndexError                        Traceback (most recent call last)
<ipython-input-5-1487e342efb9> in <module>
     11 print(x[1])
     12 print(x[2])
---> 13 print(x[3])

IndexError: list index out of range
```

The following code block demonstrates how we index elements within an array. It is possible to refer to a single element or a range of values with a colon symbol :. Therefore, we can access a slice of an array and create a subset of elements.

```
# Code Block Appendix B.6

import numpy as np

print('Working with a one-dimensional numpy array.')
x = np.array([10,20,30,40,50,60,70,80,90,100])
print(x)
print(x[0]) # first element in the array.
print(x[-1]) # last element
print(x[4:8]) # range of elements
print(x[4:]) # everything starting with the fifth element.

print('')

print('Working with a two-dimensional numpy array.')
y = np.array([[11, 12, 13],[21, 22, 23],[31, 32, 33]])
print(y)
print(y[0][2]) # element located at the first row and third column.
print(y[0,2]) # same element as the above.
print(y[1,:]) # second row.
print(y[:,1]) # second column.
```

```
Working with a one-dimensional numpy array.
[ 10  20  30  40  50  60  70  80  90 100]
10
100
[50 60 70 80]
[ 50  60  70  80  90 100]

Working with a two-dimensional numpy array.
[[11 12 13]
 [21 22 23]
 [31 32 33]]
13
13
[21 22 23]
[12 22 32]
```

APPENDIX C: PLOTS

As demonstrated throughout this book, the **matplotlib.pyplot** module is excellent for making single or multiple plots. Let us briefly summarize some essential steps in creating and managing plots.

```
# Code Block Appendix C.1

import numpy as np
import matplotlib.pyplot as plt

t = np.arange(0,2*np.pi,0.01)

plt.plot(t,np.sin(t-0.0),color='black',linestyle='solid')
plt.plot(t,np.sin(t+np.pi/2),color='black',linestyle='dotted')

plt.title('Example of single plot')
plt.legend(('first curve','second curve'),framealpha=1)
plt.ylabel('y')
plt.xlabel('x')
plt.ylim((-1.1))
plt.yticks((-1,-0.5,0,0.5,1))

plt.savefig('fig_appC_single.eps')
plt.show()
```

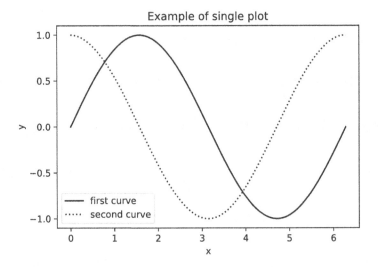

If you want to show multiple graphs, you can draw each curve one at a time within a single plot, as demonstrated in the above code block. Alternatively, you can prepare a grid of plots using **subplots()** from the **matplotlib.pyplot** module. Let us demonstrate how each subplot can be modified.

```
# Code Block Appendix C.2

import numpy as np
import matplotlib.pyplot as plt

fig, ax = plt.subplots(2,2)

t = np.arange(0,3*np.pi,0.01)

ax[0,0].plot(t,np.sin(t-0.0),color='black',linestyle='solid')
ax[0,1].plot(t,np.sin(t-0.0),color='black',linestyle='dotted')
ax[1,0].plot(t,np.sin(t-0.0),color='gray',linestyle='solid')
ax[1,1].plot(t,np.sin(t-0.0),color='gray',linestyle='dotted')

plt.subplots_adjust(left=0.1,right=0.9,top=0.9,bottom=0.1,
                    wspace=0.4,hspace=0.4)

fig.suptitle('Example of multiple plots')

ax[0,0].set_xlim((0,2))
ax[0,1].set_xlim((0,4))
ax[1,0].set_xlim((0,6))
ax[1,1].set_xlim((0,8))

ax[0,0].set_xlabel('x between 0 and 2')
ax[0,1].set_xlabel('x between 0 and 4')
ax[1,0].set_xlabel('x between 0 and 6')
ax[1,1].set_xlabel('x between 0 and 8')

# You can programmatically refer to different subplots, too.
for i in range(2):
    for j in range(2):
        ax[i,j].set_ylim((-1.2,1.2))

plt.savefig('fig_appC_subplots.eps')
plt.show()
```

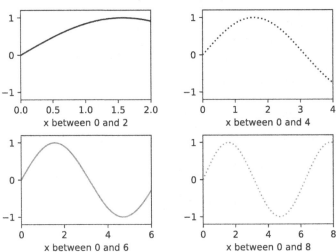

APPENDIX D: COLORS

All the graphs in the main text of this book were presented in grayscale. However, sprucing up your graphs with colors in Python is straightforward. In the **matplotlib.pyplot** module, the optional argument **color** allows you to specify colors easily by naming colors or their nicknames. For example, both **color='red'** and **color='r'** make plots in red. Over the next several code blocks, we will create a few color plots that would best be viewed on a screen rather than in print.

```
# Code Block Appendix D.1

import numpy as np
import matplotlib.pyplot as plt

t = np.arange(0,2*np.pi,0.01)

plt.plot(t,np.sin(t-0.0),color='black')
plt.plot(t,np.sin(t-0.2),color='k') # black
plt.plot(t,np.sin(t-0.4),color='red')
plt.plot(t,np.sin(t-0.6),color='r') # red
plt.plot(t,np.sin(t-0.8),color='green')
plt.plot(t,np.sin(t-1.0),color='g') # green
plt.plot(t,np.sin(t-1.2),color='blue')
plt.plot(t,np.sin(t-1.4),color='b') # blue
plt.savefig('fig_appD_color1.eps')
plt.show()
```

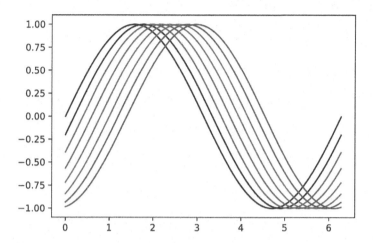

Another way of specifying colors is RGB values, a triplet corresponding to the intensities of red, green, and blue lights. For example, (1,0,0) represents red color, and (1,0,1) represents violet color, which is an equal mix of red and blue color.

```
# Code Block Appendix D.2

plt.plot(t,np.sin(t-0.0),color=(0,0,0)) # black
plt.plot(t,np.sin(t-0.2),color=(1,0,0)) # red
plt.plot(t,np.sin(t-0.4),color=(0,1,0)) # green
plt.plot(t,np.sin(t-0.6),color=(0,0,1)) # blue
plt.plot(t,np.sin(t-0.8),color=(1,1,0)) # red+green = yellow
plt.plot(t,np.sin(t-1.0),color=(0,1,1)) # green+blue = cyan
plt.plot(t,np.sin(t-1.2),color=(1,0,1)) # red+blue = violet
plt.plot(t,np.sin(t-1.4),color=(1,0,0.5)) # reddish violet
plt.plot(t,np.sin(t-1.6),color=(0.5,0,1)) # bluish violet
plt.savefig('fig_appD_color2.eps')
plt.show()
```

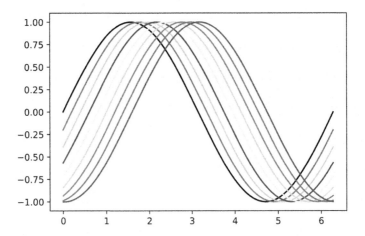

Therefore, a spectrum of gray shade can be represented by a triplet (0,0,0) (black); (0.2,0.2,0.2) (dark gray); (0.8,0.8,0.8) (light gray); (1,1,1) (white).

```
# Code Block Appendix D.3

plt.plot(t,np.sin(t-0.0),color=(0,0,0)) # black
plt.plot(t,np.sin(t-0.2),color=(0.2,0.2,0.2))
plt.plot(t,np.sin(t-0.4),color=(0.4,0.4,0.4))
plt.plot(t,np.sin(t-0.6),color=(0.6,0.6,0.6))
plt.plot(t,np.sin(t-0.8),color=(0.8,0.8,0.8))
plt.plot(t,np.sin(t-1.0),color=(1,1,1)) # white (not visible)
plt.savefig('fig_appD_color3.eps')
plt.show()
```

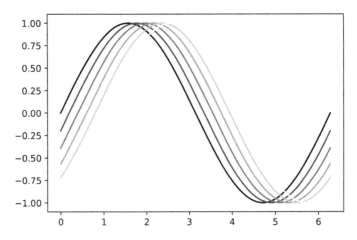

Similarly, a spectrum of red shades can be created by smoothly varying the value of red intensity in the RGB triplet.

```
# Code Block Appendix D.4

plt.plot(t,np.sin(t-0.0),color=(0.0,0,0))
plt.plot(t,np.sin(t-0.2),color=(0.2,0,0))
plt.plot(t,np.sin(t-0.4),color=(0.4,0,0))
plt.plot(t,np.sin(t-0.6),color=(0.6,0,0))
plt.plot(t,np.sin(t-0.8),color=(0.8,0,0))
plt.plot(t,np.sin(t-1.0),color=(1.0,0,0))
plt.savefig('fig_appD_color4.eps')
plt.show()
```

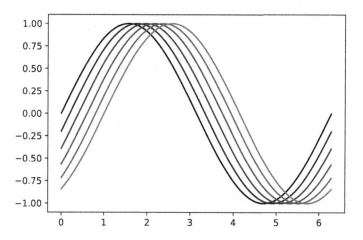

Each intensity value in the RGB triplet is often represented as an 8-bit number. Since a binary digit can take on either 0 or 1, a 2-bit number can represent four (2^2) different values: 00, 01, 10, and 11. A 3-bit number can represent eight (2^3) different values: 000, 001, 010, 011, 100, 101, 110, 111. Likewise, an 8-bit number can cover 256 distinct values. In our normal decimal system, these values go from 0 to 255.

In a somewhat unfamiliar hexadecimal system, each digit can take on 16 different symbols: 0, 1, ..., 8, 9, A, B, ..., E, F. In this hexadecimal system, a decimal value of 11 is represented as B, and a decimal value of 26 is represented as 1A. An 8-bit number (or eight-digit binary number) can be written as a two-digit hexadecimal number. For example, an 8-bit number 00000000, which is equal to 0 in regular decimal notation, can be written as a hexadecimal number 00. An 8-bit number 10000000, equivalent to 128 in decimal notation, is 80 in the hexadecimal system.

11111111 as a binary number is equal to 255 in decimal and FF in the hexadecimal system.

A notation of color you may encounter in computer programming is called hex codes, which uses the hexadecimal notation to represent RGB triplets. For example, **#FF0080** is a concatenation of **FF**, **00**, and **80**, representing 255, 0, and 128 in decimal. Therefore, this hex code is equal to (255,0,128). If we use a number between 0 and 1 in the normalized numbering system as in the previous code block, **#FF0080** corresponds to (1,0,0.5). The following code block demonstrates how these different color codes are used.

```
# Code Block Appendix D.5

# both black
plt.plot(t,np.sin(t-0.0),color=(0,0,0))
plt.plot(t,np.sin(t-0.2),color='#000000')

# both gray
plt.plot(t,np.sin(t-0.4),color=(0.5,0.5,0.5))
plt.plot(t,np.sin(t-0.6),color='#808080')

# both red
plt.plot(t,np.sin(t-0.8),color=(1,0,0))
plt.plot(t,np.sin(t-1.0),color='#FF0000')

plt.savefig('fig_appD_color5.eps')
plt.show()
```

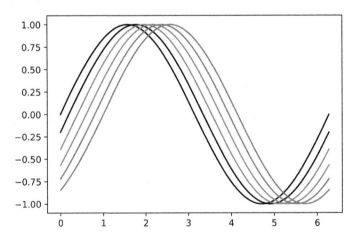

APPENDIX E: ANIMATION

The collision process in Chapter 2 can be nicely visualized with an animation. As a warm-up, let's create a sketch of one gas particle inside a container. We will specify the position of a particle with variables **x** and **y**, and use the **scatter** command from the **matplotlib** module. We will use the **plot** command to draw thick vertical lines representing the wall. For mathematical simplicity, we will only consider one-dimensional, vertical up-or-down motion. Study the following lines of code. Experiment with the code, so that you understand what each line does. For example, can you change the color of the walls? Can you make the wall thicker? Can you move the position of the particle?

```python
# Code Block Appendix E.1
import matplotlib.pyplot as plt

# Place the particle.
x, y = (0.4, 0.5) # position of a particle.
plt.scatter(x,y)
plt.text(x+0.05,y-0.05,'Particle at (%2.1f, %2.1f)'%(x,y),
         color='red')

# Draw the walls, which are located at 0 and 1.
plt.plot((-0.1,1.1),(0,0),color='black',linewidth=5)
plt.plot((-0.1,1.1),(1,1),color='black',linewidth=5)
plt.xlim((-0.1,1.1))
plt.ylim((-0.1,1.1))

plt.axis('off')
plt.savefig('fig_appE_particle_in_box.eps')
plt.show()
```

● Particle at (0.4, 0.5)

The following code block makes an animation of a moving particle between two walls, using a Python module `matplotlib.animation`. The animation routine relies on the previously-developed function `calculate_position()`, which tracks the position of a particle at different times.

We are also utilizing the `plt.subplots()` command, which returns the figure and axis attributes (`fig` and `ax`) of a generated plot, so that they can be manipulated easily. For instance, `ax.cla()` clears the current data points from the axis, so that the new set of data points can be drawn for the next frame in the animation.

Feel free to play around with the code by trying out different parameters. For example, you can change the initial position and velocity of a particle represented by the variables, `x0, y0, v`.

```
# Code Block Appendix E.2

# Animate the position.
import numpy as np
import matplotlib.animation as animation
from matplotlib import rc
rc('animation', html='jshtml')

def calculate_position (y0,v,ymin=0,ymax=1,dt=0.01,
                        tmin=0,tmax=10,plot=False):
    # ymin and ymax are the boundaries of motion (walls).
    current_v = v
    time_range = np.arange(tmin,tmax,dt)
    y = np.zeros(len(time_range))
    y[0] = y0

    Nbounce = 0
    for i, t in enumerate(time_range[1:]):
        current_y = y[i] + current_v*dt # Update position.
        if current_y <= ymin:
            # if the particle hits the bottom wall.
            current_v = -current_v # velocity changes the sign.
            current_y = ymin + (ymin - current_y)
            Nbounce = Nbounce+1
        if current_y >= ymax:
            # if the particle hits the top wall.
            current_v = -current_v # velocity changes the sign.
            current_y = ymax - (current_y - ymax)
            Nbounce = Nbounce+1
        y[i+1] = current_y
    if (plot):
```

```
        plt.plot(time_range,y)
        plt.xlabel('Time')
        plt.ylabel('Position')
        plt.savefig('fig_ch2_bounce.eps')
        plt.show()
    return y, time_range, Nbounce

fig, ax = plt.subplots()
x0 = 0.3
y0 = 0.5
v = -0.2
position, _, _ = calculate_position(y0,v,dt=0.5,tmax=20)

def animate(i):
    ax.cla()
    plt.scatter(x0,position[i])
    # Draw walls
    plt.plot((-0.1,1.1),(0,0),color='black')
    plt.plot((-0.1,1.1),(1,1),color='black')
    plt.xlim((-0.1,1.1))
    plt.ylim((-0.1,1.1))
    plt.axis('off')

ani = animation.FuncAnimation(fig, animate,
                            interval=50, frames=len(position),
                            repeat=False)
ani
```

The above code block will produce an animation of a single particle
moving up and down. In the following code block, we animate the motions of N particles that are evenly spaced out.

```
# Code Block Appendix E.3

# Animate multiple particles.
# Note this code may take some time to complete.

N = 30
tmin = 0
tmax = 10
dt = 0.1
t = np.arange(tmin,tmax,dt)
pos = np.zeros((N,len(t))) # initialize the matrix.
Nbounce = np.zeros(N)

v = np.random.randn(N)*0.5
y0 = np.random.rand(N)
for i in range(N):
```

```
        # pos[i,:] references the i-th row of the array, pos.
        # That is the position of i-th particle at all timestamps.
        pos[i,:], _, Nbounce[i] = calculate_position(y0[i],v[i],dt=dt,
                                                      tmin=tmin,tmax=tmax)

fig, ax = plt.subplots()

def animate_Nparticles(i):
    ax.cla()
    N, frames = pos.shape
    x = np.linspace(0,1,N)
    for j in range(N):
        plt.scatter(x[j],pos[j,i],color='gray')
    # Draw walls
    plt.plot((-0.1,1.1),(0,0),color='black')
    plt.plot((-0.1,1.1),(1,1),color='black')
    plt.xlim((-0.1,1.1))
    plt.ylim((-0.1,1.1))
    plt.axis('off')

N, frames = pos.shape
ani = animation.FuncAnimation(fig,animate_Nparticles,
                              interval=50,frames=frames,
                              repeat=False)
ani
```

Epilogue

A world-renowned physicist[†] said about thermal physics that "it is the only physical theory of universal content concerning which I am convinced that, within the framework of applicability of its basic concepts, it will never be overthrown." Indeed, the concepts and frameworks of thermal physics are being used to study such wide-ranging topics as black holes, communication, computation, biological phenomena, artificial intelligence, and the nature of reality ("it from bit"). We enjoyed writing this book and hope the readers found our presentation of thermal physics enjoyable, too.

[†]See "Thermodynamics in Einstein's Thought" by Martin Klein in Science Vol. 157, No. 3788 (1967).

Index

Printed in the United States
by Baker & Taylor Publisher Services